AF606906

Praise for *Lost in Curiosity*

"*Lost in Curiosity* takes readers into the laboratory and out into the field to explore science as a work in progress. Roberta Kwok has written a lively and revelatory book, and, as science in the U.S. increasingly comes under attack, it couldn't be more timely."

—Elizabeth Kolbert, Pulitzer Prize–winning author of *The Sixth Extinction: An Unnatural History*

"Opening *Lost in Curiosity* is like walking through a door into a sometimes strange but always beautiful land, full of the unexpected adventures that science offers, the puzzles, the challenges, and the bright flash of understanding when it all comes together. It's the kind of trip you won't want to miss."

—Deborah Blum, Pulitzer Prize–winning author of *The Poison Squad: One Chemist's Single-Minded Crusade for Food Safety at the Turn of the Twentieth Century*

"*Lost in Curiosity* shows that the doing of science is not a tidy affair with clear answers following in a straight line from clear questions. Instead, there are wrong turns, misguided assumptions, broken equipment, and a great deal of thrashing around… Both colorful and beautifully written, Kwok's book tells not only this story but also authentically conveys the daily life and mind of the scientist."

—Alan Lightman, coauthor of *The Shape of Wonder: How Scientists Think, Work, and Live*

"*Lost in Curiosity* tackles my favorite part of the scientific process:

when experiments go moldy and frogs refuse to be found and malfunctioning equipment conjures moons out of thin air. Roberta Kwok is a trustworthy guide as she shares tantalizing peeks into the murk that precedes discovery and proves that our instinct to do science, and better understand our world, is what makes us human. One of the most honest books about science I've ever read, and an utter delight."

—Sabrina Imbler, author of *How Far the Light Reaches: A Life in Ten Sea Creatures*

"It's a rare book that finds thrills not in the triumph of discovery but the messiness of the journey. Roberta Kwok's warm, humane portraits of scientists in the midst of unfinished work jump from jungles to deserts and span the cosmic to the microscopic. *Lost in Curiosity* is an enthralling glimpse of hidden worlds: delightful, bright with insight, and deeply necessary."

—Melissa L. Sevigny, author of *Brave the Wild River: The Untold Story of Two Women Who Mapped the Botany of the Grand Canyon*

"From Borneo to Greenland to Jupiter's moons, *Lost in Curiosity* is an engaging, globe-spanning glimpse into the lives of everyday field scientists. Roberta Kwok expertly details the passion and chance that ignite scientific inquiry, and reveals how frustration and struggle go hand in hand with discovery."

—Jonathan C. Slaght, award-winning author of *Tigers Between Empires: The Improbable Return of Great Cats to the Forests of Russia and China*

"*Lost in Curiosity* is a fascinating exploration of the process of science and the scientists who pursue it. From studies that failed to those that raised more questions than they answered, Kwok deftly threads the twists and turns that define the scientific enterprise… Readers will come away feeling closer to those who have chosen science—and more than a little in awe of their tenacity."

—Bethany Brookshire, author of *Pests: How Humans Create Animal Villains*

"Insightful and beautifully written, *Lost in Curiosity* tells the stories of scientists at work as they grapple with the 'messy middle' of science—the tricky, unpredictable work of carrying out experiments. Kwok takes readers into the daily lives of scientists as they deal with equipment failures, experiments that go awry, and carefully laid plans derailed… Kwok's gorgeous writing makes her subjects feel alive on the page, and through their stories, she convincingly argues that scientific work is worth doing even when the outcome is uncertain."

—Christie Aschwanden, *New York Times* bestselling author of *Good to Go: What the Athlete in All of Us Can Learn from the Strange Science of Recovery* and producer/host of the *Uncertain* podcast

"*Lost in Curiosity* is a charming, compelling, and much-needed reminder that even the least glamorous work can be full of beauty, and even dead ends tell a larger story."

—Michelle Nijhuis, author of *Beloved Beasts: Fighting for Life in an Age of Extinction*

Lost in Curiosity

Lost in Curiosity

FIELD NOTES FROM SCIENTISTS' ADVENTURES INTO THE UNKNOWN

Roberta Kwok

Cover design by Caitlin Sacks
Cover images © B.illustrations/Shutterstock, Gringoann.art/Shutterstock, Ideacartoons/Shutterstock, Melodiana Studio/Shutterstock, Alona K/Shutterstock, enesdigital/Shutterstock, 3d_kot/Shutterstock, ArtCreationsDesignPhoto/Shutterstock, Mantella Aurantiaca/Shutterstock, Pani Malyuvalka/Shutterstock
Internal design by Tara Jaggers/Sourcebooks
Internal art by Emma Regnier

Parts of the introduction originally appeared in different form in the blog post "'Through a Glass Darkly': How Science Seeks Truths in Hidden Worlds" in June 2021 on the MIT Knight Science Journalism Program website.

Parts of the evolutionary biology chapter originally appeared in different form in the article "Devoted Dads of the Amphibian World" in December 2018 on the *New York Times* website.

Published by Sourcebooks
1935 Brookdale RD, Naperville, IL 60563-2773
(630) 961-3900
sourcebooks.com

Cataloging-in-Publication Data is on file with the Library of Congress.

Printed and bound in the United States of America.
VP 10 9 8 7 6 5 4 3 2 1

For Gillian

We are creatures of constant awe,
curious at beauty, at leaf and blossom,
at grief and pleasure, sun and shadow.

—Ada Limón, "In Praise of Mystery: A Poem for Europa"

Contents

Introduction

Into the Wardrobe

"F*uck.*"

It was a Friday night, shortly after 11 p.m., in winter 2006. I was standing in a small room full of plants in the biology department of Indiana University Bloomington, a college about an hour south of Indianapolis. The campus had emptied out; people were knocking back drinks at bars, ensconced at the movies, or asleep at home after a long week of work.

The person who had just uttered the expletive was my friend Ben Blackman, a PhD student who was studying how sunflowers enter "puberty" and start flowering. I was an English graduate student at the time, and as research for a science essay, I'd been tagging along with Ben on a twenty-hour experiment. Starting early that morning, we'd snipped about seventy leaf samples every four hours and tucked them into envelopes for Ben to analyze later.

Things had started going wrong almost immediately. At 7 a.m., Ben banged his head on a shelf and knocked over a flowerpot, spilling dirt on the floor. Around 7 p.m., he realized that we'd put several dozen leaves in the wrong envelopes, and we had to relabel all of them. "It's my fault," he said. "I should have checked them." (Looking back, I'm pretty sure I was the one responsible for this snafu.) He'd run into troubles earlier too. One growth chamber was malfunctioning, so about a week ago he'd had to move some plants to another lab's room.

Now, close to midnight, Ben had found out that the new room's light configuration was wrong. One critical factor in his experiment was the number of hours of light exposure the sunflowers received each day; by 11 p.m., the room should have gone dark automatically. But it was still bright, and the control panel display said the lights weren't programmed to go off for another two hours. Since the plants had already been growing there for a week, his data might be worthless. It was the first time that day that he seemed genuinely upset.

After we took the final leaf samples at 3 a.m. and found the light programming still acting wonky, Ben decided that he'd probably need to redo the entire experiment. But he regained his usual cheeriness, saying he had other tasks he had to stay up late for anyway. I asked how this experiment ranked in terms of how well it went, what with the growth chamber malfunction, the spilled flower pot, the mixed-up envelopes, and the botched light program.

"Oh," he considered. "About average."

The word "science" often brings to mind the image of a lone genius struck with a brilliant idea: a flash of insight, a thunderclap of clarity. But the reality is far more muddled. Researchers spend most of their time mired in the "messy middle" of science—the opposite of a Eureka moment. Most can relate to Ben's tribulations, the obstacles and indignities varying depending on their field of study; instead of misconfigured lights, they might be subjected to blistering blizzards, constant poison oak rashes, baffling telescope data, or elusive wild creatures that handily evade humans' earnest attempts to observe them. "Eureka!" means "I have found it!" but a more fitting motto for science might be "I can't find it," "Where is it?" or, most often, "What the hell is going on?"

As a former (let's be honest, failed) researcher, I know all too well that the scientific method doesn't always go as planned. I studied biology in college and joined cancer and immunology labs, intending to pursue research as my career. However, I soon realized that I had neither the talent for processing samples precisely or fixing temperamental instruments, nor any tolerance for the tedium of repeating the same analyses over and over. After a few years of frustration, I put down my pipettes and became a software engineer, and then a writer. The sunflower study with Ben, several years after I'd left the lab, was the first time I'd seen the process up close since my own ill-fated endeavors; the experience brought back memories of times when an uncalibrated instrument, an incorrectly categorized tube, or an unwitting miscalculation could bring an entire experiment crashing down.

Eventually, I became a science journalist, reporting on topics ranging from astronomy to zoology. My articles had a typical format:

A study was completed, a report published, a question answered. But I knew that my quick interviews with scientists and tidy news headlines didn't capture the months or years of failure, doubt, and uncertainty that underpinned each hard-won result. I felt like I was missing the best parts of the stories—like finding out that Frodo and Sam had delivered the Ring to Mount Doom without hearing about their harrowing journey from the Shire, or that the White Witch had finally been defeated in Narnia, without knowing how Lucy found the magical entrance in the wardrobe, Edmund betrayed his siblings for Turkish Delight, and Peter slew a giant wolf.

I wrote this book so that I could tell those hidden stories. For each chapter, I followed scientists who were caught in the middle of their quests. And reporting the book was indeed like entering a series of wardrobes, each leading to a different world. I ended up in the Arizona desert, at the edge of an abandoned uranium mine; at a wave laboratory in Oregon, wading in a tank of soggy, slightly smelly "tutus" intended to help protect cities from storms; on a remote glacier in Greenland, flying over a huge crevasse that gashed the ice; at a museum in Washington, watching the hair-raising (and also smelly) dissection of coyote carcasses; and traveling, in my imagination, more than 600 light-years from Earth to a volcanic moon spewing fountains of lava. Sometimes I got utterly lost—usually when I was sitting in on meetings with physicists, as they held spirited discussions of hexagonal meshes, polydisperse particles, quadratic ramping, and something mysteriously named the Warren-girder case (which, one of them told me, was "an absolute pain in the ass").

Along the way, I saw the tiny details that scientists had to pursue to answer big questions. Though many of the researchers grappled with lofty ideas or critical global issues, their labor behind the scenes was often humble. To uncover part of the history of human evolution, a paleontologist counted hundreds of faint lines on a fossil; to help retrace COVID-19 chains of infection, biologists spent endless days sequencing virus samples, each DNA strand tugged through microscopic pores on a machine like fine thread through embroidery cloth. As philosopher of science Michael Strevens wrote in his book *The Knowledge Machine*, "it is by accounting for the fraction of an inch, the sliver of a degree, that the truth will make itself known."

This book ultimately isn't a formal treatise on the process of science. It's the oldest and simplest kind of book: a storybook. Some chapters have a clear protagonist, while others have a sprawling international cast. Some have happy endings, some sad, and some bittersweet. (Spoiler alert: They're mostly bittersweet.) But the researchers' stories never truly end. What I've recorded is a leg of each team's journey, a fleeting glimpse into their lives as I joined for the trek between two waypoints and then waved goodbye as the scientists ventured on into the hazy distance. The stories are similar to what poet Stanley Plumly once said about poems: "They begin in the middle and they end in the middle, only later."

To select the studies in this book, out of the thousands I could have followed, I relied on an admittedly unscientific method. I chose them because when the researchers told me about the questions they planned to investigate, I wanted to know what would happen next. In my mind, I loosely categorized the scientists as "fixers" and "dreamers": those who were hurling themselves at the seemingly intractable problems of our time, such as climate change or environmental pollution, and those pursuing a deeper understanding of nature and the universe, seeking the untold stories of intriguing creatures, faraway planets, or even everyday objects. What they had in common was curiosity, persistence, passion; a desire to explore and make sense of our world; a largeness of perspective, but also a tender attention to the small, the vulnerable, the neglected, the misunderstood. And by the end, their narratives entwined in subtle ways that I didn't expect. From the initial randomness emerged a kind of order.

Though my time with the scientists was short, some moments have stayed with me. One occurred when I visited a paleontology lab in Seattle, close to my hometown of Kirkland, Washington. On a computer screen, I saw a magnified image of a fossil. The scientist turned on a polarized light filter, and flecks of red, azure, gold, violet, pink, and tangerine appeared, as if sunlight had shone through a stained-glass window. Later, another researcher told me that paleontologists see the objects of their studies "but through a glass darkly," an echo of an oft-quoted passage from the Bible. This, I imagine, is what it's like to do science: straining to see the truth through panes of colored glass. The tortured process never yields perfectly clear answers. But there's something beautiful about it too, isn't there?

Helheim Glacier

GEOLOGY

The Invisible River

"We're doing something. We're doing science. I don't care if it's a terrible approximation of what we set out to do. But we're putting stuff in the ground and it's happening."

Jessica Mejia was shaking her head.

It was June 6, 2022, six weeks before she was supposed to leave for an expedition in Greenland. Jess, a postdoctoral researcher in glaciology at the University at Buffalo in New York, had planned to spend about ten days setting up instruments on the southeastern part of the ice sheet. Her team would camp on Helheim Glacier, named for a Norse world of the dead—depicted in the legends as a shadowy, foggy realm, where a giantess presided over gated palaces and a river littered with swords. Jess wanted to understand the hidden world under the ice.

I'd been following her work for the last year and a half, and she'd told me that she hoped to study the flow of melted water into crevasses to the bottom of the ice sheet—a mysterious process, invisible from the surface, that might influence sea-level rise. Now, on a video call about her team's progress, I asked, "Is it good or bad?"

"It's not good," she said. "Everything is a nightmare with the helicopter."

Jess had been preparing for months. She'd assembled shovels, gloves, wire cutters, medical kits, sleds, bungee cords, batteries, drill bits, wrenches, duct tape, and dozens of other items. With her teammates, Jess had tried to anticipate all the problems that could arise, discussing everything from whether they needed specialized jerry cans for fuel to the legal ramifications if they had to shoot a hungry polar bear that wandered into camp. Her collaborators in New Hampshire had packed a shipping container of supplies, including hundreds of pounds of food: rice, beans, pasta, mac and cheese, soup, ramen, lots of chocolate.

Now they'd lost their booking for the helicopter, which was supposed to fly the team from Tasiilaq, a coastal town in Greenland, to their field site about sixty-five miles inland on the glacier. Supply chain issues had prevented the company from getting the proper gear; helicopter time had also gotten competitive because of surging interest from tourists and mineral exploration firms. The team could go in September instead, but they would face a higher risk of fog and storms. If visibility was poor, they could get stuck in town for days.

"My big concern is that the weather will just be absolute crap, and we won't be able to get on the ice," Jess said.

This wasn't the first time she'd run into fieldwork obstacles. In 2018, her team had been delayed in town for a week, waiting for the weather to clear so the helicopter could fly to their research site in western Greenland. The year before that, a snowstorm hit while they were camping and confined them for several days to their tents, where Jess heard ice hitting the fabric all night and watched hours of *RuPaul's Drag Race* episodes on her phone. And her current team's fieldwork had already been postponed for two years because of the COVID-19 pandemic.

But Jess was not easily quashed. She had a no-nonsense, rapid-fire way of speaking and seemingly endless energy for solving problems. Growing up in Connecticut, she'd wanted to become a lawyer and the first female president. As a teenager, she struggled with big questions—*Why are we here? Is there a god? What's the point of everything?*—and thought she'd study cosmology. In college, though, physics research projects seemed too abstract, either observing something far away in space or manipulating materials at a tiny scale. She wanted something she could experience, something *real.*

Then she came across beautiful pictures of glaciers online, in which researchers had used pink dye to trace the flow of water. She read that the field had huge unanswered questions, and scientists needed to figure out how fast glaciers would lose ice and contribute to sea-level rise. Those critical gaps in knowledge gave her the reason she needed to pursue research, the answer to her teenage questions about the point of it all. She entered a graduate program in geophysics, focusing on glaciology, and became fascinated by what was happening under the ice. On her first expedition to Greenland, in her tent at night with her ear to the ground, she could hear the glacier's deep, echoey cracking. And at the edges of crevasses, she could hear water trickling, dripping, sometimes thundering inside.

I would meet her in person in Greenland about a year later. In one of the most emblematic photos I took, Jess was carrying gear to the shipping container, her long dark hair under a Mets baseball cap, a coil of neon green rope in one hand and, slung over her shoulder, a giant black mallet that her team had likened to the Mario hammer. What Jess did day to day wasn't so different from Mario's trials: always in motion, jumping over obstacles, collecting whatever she could along the way, and trying to get to the next stage.

After the summer helicopter booking got canceled, her team decided that they should go in September and do whatever they could. The project kept dragging on; they needed to get instruments on the glacier so they could collect data over the next couple of years. When I texted Jess the day before her departure, she said she was feeling good, just packing and taking care of last-minute details. Once she was camping on the ice, I figured I wouldn't hear much from her. But on a Sunday morning around 5:30 a.m. my time, when she was still supposed to be on the glacier, I got a message from Jess: "I have a long story for you."

Helheim Glacier was an unforgiving place. This massive blanket of ice, more than half a mile thick, lay over mountain peaks and deep canyons; the resulting landscape was a silent, slightly undulating ocean of white. Scientists told me that when they camped on the glacier, the only wildlife they saw were small songbirds called snow buntings, and the only sound was the wind blowing.

In 2011, researchers had discovered something completely unexpected beneath the ice. Clément Miège and Evan Burgess, then graduate students at the University of Utah in Salt Lake City, had traveled to Helheim Glacier with a mountaineer and field engineer to measure how much snow fell each year. The researchers planned to drill about 200 feet down to collect cores, columns of old snow and ice that they'd analyze in the lab. It was a slow, tedious, repetitive process, with no surprises.

One day, when the engineer pulled up a core that had gone down about thirty feet, water spilled and dripped from the drill. Temperatures had hovered around –50 to –30°F over the last month, and even though that day was a relatively balmy 5 degrees, it still wasn't nearly warm enough for snow to melt. Burgess waved at Miège, who was working nearby, and said, "We hit water! There's water in the snow." When they called their adviser, Rick Forster, by satellite phone, he was stunned. "To see liquid water gushing out of a core is just incredible and shocking," said Forster, a glaciologist at the University of Utah.

Burgess and Miège had uncovered the first evidence of a giant aquifer hidden in the ice. Later measurements suggested that the aquifer covered more than 800 square miles and extended an average of about ninety feet below the surface. Radar data revealed signs of aquifers in other parts of Greenland too, adding up to a total area roughly the size of Massachusetts. The water was stored in firn—old snow that hadn't completely compacted into ice and had a lot of gaps, which acted like a giant sponge. Snow insulated the aquifers from cold temperatures like a blanket and prevented the water from refreezing.

The discovery raised questions about how firn aquifers could affect sea-level rise. More water was likely seeping into the aquifers every year when snow melted on the surface in the summers. Did the aquifers trap the water for decades, acting as a buffer and slowing down sea-level rise? Would the water eventually refreeze, or would it drain to the bottom of the glacier? If the latter, would the water trickle slowly and steadily or gush out in a catastrophic flood?

Kristin Poinar, a glaciologist now at the University at Buffalo, wondered what would happen when crevasses intersected with the aquifer. If those crevasses filled with water, the pressure could force the cracks to open deeper, fill with more water, and fracture all the way to the underlying bedrock. The water would then have a clear path to reach the ocean. In a 2017 study, her team's computer model suggested that the aquifer water could make crevasses crack to the bed in ten days to six months.

Jess, who had joined Kristin's lab in 2021, wanted to observe this process in detail on the ice sheet. Her plan was to install GPS stations on both sides of crevasses on Helheim; if they moved apart gradually, aquifer water was probably filling the crack and making it wider and deeper. If the units' movement stopped, the crevasse had likely fractured to the bed. Meanwhile, Jess's colleagues would use other methods, such as radar and seismic signals, to track water through the ice and take more measurements of the aquifer's volume. With this data, scientists could better understand how much water was reaching the bottom of the glacier over time.

Their ultimate goal was to improve predictions of sea-level rise, which at the moment are alarmingly vague. Scientists' long-term projections have huge error bars, with estimates anywhere from about six to twenty-two feet—and perhaps even more than forty-nine feet—by the year 2300 under a very high emissions scenario. Researchers don't know how positive feedback loops will play out—that is, how melting will trigger other processes that generate more ice melt or loss. For instance, large-scale ice sheet models don't fully account for the effects of water flowing through the Greenland ice sheet to the bed. They assume a steady, constant melting, Forster said. But "these aren't just ice cubes that are slowly melting. They're much more complicated than that."

The effect of Greenland's firn aquifers draining into the ocean wouldn't make a big difference to global sea levels. But because Greenland is warmer than Antarctica, scientists could use what they learn there to predict Antarctica's fate in future centuries. If the entire Greenland ice sheet melted, sea level would rise by about 24 feet; for Antarctica, that figure would be an almost unthinkable 190 feet, driving much of the uncertainty around long-term projections. In 2018, researchers confirmed the presence of an aquifer in Antarctica for the first time, and more will likely appear in the future, said Michiel van den Broeke, a polar climatologist at Utrecht University in the Netherlands. Firn aquifers form when a large volume of snow melts on the surface in summer, and then a lot of snow falls in autumn and winter to insulate that water from refreezing. "That's exactly what we think is going to happen," he said. "It will also create the perfect conditions for these aquifers to form."

The repercussions in Antarctica could be huge, said Amber Leeson, a glaciologist at Lancaster University in the UK. If the water in firn aquifers drives crevasses all the way through ice shelves, the fractures could make ice shelves collapse. And the disintegration of one critical ice shelf could cause a few feet of global sea-level rise.

When I asked van den Broeke how the work by Jess's team fit into the big picture, he told me that it was extremely important to gather this type of data to fuel computer models that forecast sea-level rise. He said, "It is this research, this detailed research, that we really need to be able to make any prediction at all."

The day after Jess sent me that message from Greenland, we got on the phone. I asked, with some trepidation, "What happened?"

The field season had been, to put it bluntly, a disaster. The expedition had been a joint effort by three groups: Jess's team from the University at Buffalo, another from Dartmouth College in Hanover, New Hampshire, and a third from Georgia Tech in Atlanta. Eight scientists had gone on the trip, with two mountain guides to oversee safety. The plan was for Jess and the guides to fly from the town of Tasiilaq to the glacier first, followed by the rest of the researchers and the gear.

But when the team arrived in Tasiilaq and Jess saw the helicopter, she felt alarmed. It was smaller than any she'd ever worked with; she was supposed to deploy several GPS units, each packaged with four car battery–sized batteries in a case about as big as a mini-fridge, and there was no way that even one case would fit in the cabin. The helicopter could transport sling loads in a net attached with a cable, like a stork carrying a baby, but the pilot's ability to sling would depend on weather. Passengers wouldn't be allowed on sling-load flights, and Jess worried that the team would run out of time to get all the researchers and gear onto the ice. The batteries for one GPS unit alone weighed more than 250 pounds. And her team was splitting the use of the helicopter with another research group, limiting their flights.

After waiting for the fog to lift on the first flight day, Jess and the guides managed to get to the glacier field site by around 6 p.m. But the next morning in town, Renée Clavette, a graduate student from the Georgia Tech lab, woke up feeling like she had a fever. When she took a COVID test, the red positive line showed up immediately.

At this point, more problems started to topple their carefully laid plans like dominoes. Fog kept delaying further flights. Four of Jess's teammates made it to the ice but had to leave the GPS units and other equipment behind. The pilot wasn't willing to carry a sling load and had to leave town unexpectedly. Then the inevitable happened: Jess and Aleah Sommers, a postdoctoral researcher at Dartmouth, tested positive for COVID at camp. Soon afterward, Jess was overcome with nausea and couldn't eat more than a couple Clif Bars, a bite of a peanut butter and jelly sandwich, and a bit of a cracker.

The team's worries ratcheted up. They now wanted the helicopter to transport at least one of the remaining researchers and the science gear to the field site and to take the two COVID patients back to town. They also needed more fuel for cooking at camp. At the beginning of the trip, logistical delays had compressed their schedule for organizing equipment, and in the chaos of unpacking and rearranging gear to fit in the helicopter, they hadn't brought enough white gas for the stoves. But the helicopter never made it to their camp that day for one reason or another—fog in Tasiilaq, the replacement pilot unable to get into town. Then in the evening, the team found out that a storm was coming.

The guides, Mike Coyle and Richard Mansfield, took a step back and assessed the risks. Their margin for error was shrinking. The camp was in an incredibly remote location, and they depended completely on the helicopter. Imagine that you sailed far out in the ocean in a small boat, said Mike, a guide from Colorado Mountain School in Boulder. Then the water froze around you, and then two people on the boat got sick, and one wasn't eating, and your only possible rescue transport couldn't reach you because of bad weather. Richard, the founder of Mountain Adventure Guides in Devon, UK, told me in his understated way that the situation "just wasn't ideal." If Jess's or Aleah's condition worsened and the helicopter didn't pick them up, they'd be stuck on the ice in the storm without medical care. Later, Jess said that this is how tragedies happen. One thing goes wrong, then two, three, four—and then there's a tipping point.

The team's mindset shifted: Instead of getting more people on the ice, they needed to get everyone off. And given the number of people at camp to transport, they needed to start flying back the next day.

At this point, only three scientists were left standing. They had come on the expedition with ambitious plans: Colin Meyer, a glacial fluid dynamicist at Dartmouth, and his team wanted to measure the snow's density and how easily water could flow through it. Winnie Chu, a glacial geophysicist from Georgia Tech, was going to install radar instruments that gave a two-dimensional view of the glacier all the way to the bed, like an ultrasound. And Courtney Shafer, a PhD student in Kristin's lab, had hoped to measure the top and bottom levels of the aquifer by smacking the ice with a mallet or setting off a small explosion to create seismic waves.

When the situation was looking dire, Colin had tried to rally the troops. "Guys, we're doing something," he said. "We're doing science. I don't care if it's a terrible approximation of what we set out to do. But we're putting stuff in the ground and it's happening." Other team members called Colin "the Energizer Bunny"; a frequent high-fiver with a bit of a swagger, he positively appeared to bounce when he walked, and he was already so perky in the mornings that he drank decaf English breakfast tea at camp instead of coffee. Over the next couple of days, the researchers installed one radar unit, dug snow pits, and tried to generate seismic signals, Colin whacking the ground as hard as he could with a rubber mallet. The wind picked up, and the helicopter pilot, after flying to the edge of the ice sheet, decided it was too dangerous to continue toward the field site.

In the end, the story was anticlimactic in the way that stories of safe, good decisions are. The wind died down, everyone returned to town before the storm arrived, and Jess and Aleah recovered. The following week, Winnie and two team members flew back to the glacier to install a second radar unit—one of the expedition's few successes.

Jess was matter-of-fact about how the trip had turned out. "Not as planned," she said. Colin was less diplomatic. It was like "we made it to the championship game," he said, "and then we got our asses kicked."

Aleah had a philosophical take. Spirituality was an important part of her life, and when their first helicopter booking had gotten canceled, she told me that she tried to appreciate the moment as something akin to a Tibetan mandala. A mandala was an elaborate diagram meticulously filled with colored sand grains such as crushed marble—and once it was complete, you swept it away. It was a way of valuing the process over the results. They'd accomplished some of what they'd hoped, she said, and "you let the rest go for now."

Nine months later, in June 2023, I took a red-eye from Seattle to Reykjavík, Iceland, and then boarded a plane to Greenland. We would fly about two hours across the Denmark Strait to Kulusuk, a small town on an island off Greenland's southeast coast. From there, I'd take a fifteen-minute helicopter ride to Tasiilaq, a town of about eighteen hundred people on another nearby island, where I'd meet Jess's team. Though I wouldn't be able to stay on the ice with them—they'd had a hard enough time getting all the scientists to the field site the previous year, let alone a journalist—I might take a quick day trip to camp and could follow their progress from town via satellite calls.

The researchers' preparations had gone well: They'd hired a different helicopter company, and the weather in June was supposed to be better. The team on the ice was also smaller: Jess, Colin, Winnie, Mike, Richard, and two new graduate students. For the GPS stations, Jess planned to identify a safe path to ski to the crevasses after arriving at camp, and then have a pilot drop off the instruments nearby. Before leaving, I'd asked if she was nervous about delaying the sling load of crevasse GPS units until after they'd mapped a path. "Yes," she said. "*Yes.*"

Most of the other people on my flight appeared to be members of a tour group, retirees in outdoor gear and hiking boots. Out the window, the clouds thickened and thinned; then specks of white began to appear far below. As we approached Greenland, I saw giant swirls of sea ice fragments and icebergs scattered over the water; they reminded me of the Milky Way, that same feeling of a vast spiraling of matter.

Ice sheets might look like solid, rigid blocks, but they behave like a viscous, slowly moving fluid. Greenland is essentially a dome, colder at the high-elevation interior and warmer near the edges. When snow falls on the middle of the dome and turns into ice, the weight makes ice flow toward the coasts. Imagine pouring very thick pancake batter in the center of a pan and seeing it spread outward, said Amber Leeson, the glaciologist at Lancaster University. If you stuck a flag onto Jess's field site on Helheim Glacier and left it for a year, it would move with the ice more than 600 feet. And if you stood on the ice sheet, you would hear loud creaking and groaning, as if you were standing in an old house.

For sea-level rise projections, the critical question is: How quickly are Earth's ice sheets losing mass? If you poured pancake batter onto a slightly dome-shaped pan with unrimmed edges, bits of it would drip off onto the stove; similarly, when ice flows toward the coast, icebergs calve into the ocean. The quicker the ice moves, the more icebergs drop off. In Greenland, as the ice flows to warmer, low-elevation areas, the surface also melts into water and runs into the ocean. But ice sheets can offset those losses by gaining new ice during the winters through snowfall; at Jess's field site on Helheim, for instance, about thirteen feet of snow can fall each year.

In the 1990s, scientists didn't know which process was winning—that is, whether Greenland and Antarctica were losing more mass than they were gaining. "We simply had no clue," said van den Broeke, the polar climatologist at Utrecht University. But researchers generally thought the ice sheets would react very slowly to changing climate and that "they are too big to fail," he said.

Then in the mid-2000s, satellite data showed that Greenland and the West Antarctic Ice Sheet were already losing mass. "Suddenly we realized it is happening now," van den Broeke said.

"It is ongoing as we speak." In Antarctica, warmer ocean water was melting the ice shelves from below; the thinning and weakening of these shelves enabled more ice to flow from the interior to the edges, which in turn made more icebergs break off. And over the next couple of decades, Greenland's ice continued to take researchers by surprise. During the summer of 2012—which one scientist called a "huge, jaw-dropping" record event—nearly the entire ice sheet melted on the surface, even at its summit. Seven years later, the summit melted again.

That meltwater could have other insidious effects. Glaciologists had assumed that water on the surface of the ice sheet didn't affect how quickly the ice flowed outward. But in 2002, NASA scientists and their colleagues showed that liquid water likely drained to the bedrock and lubricated the base of the ice sheet. That layer of water underneath seemed to make the ice flow faster toward the coasts, as if the pancake batter were sliding over a slick of oil.

Some scientists wondered if this phenomenon could generate a runaway effect: As temperatures rose, more water would melt on the surface and reach the bed, making more ice flow outward to warmer, lower elevation areas, where it would melt even more, and so on. Winnie Chu recalled that when she was starting her PhD program, "everybody's freaking out," thinking "whoa, if that is the case, then if we have two degrees warming and then we generate tons more water, that means all these glaciers are going to slide into the ocean."

The more I learned about the effect of meltwater, though, the more confusing the picture got. On the one hand, turbulent water at the bottom of the ice sheet could melt open tunnels, which would let the water flow through without lubricating the ice. On the other hand, fresh cold water emerging from the base of the ice sheet into a fjord could make warmer water circulate, melting the edges of the ice sheet faster. But maybe the firn aquifer could hold a large reserve of water and prevent it from reaching the bottom for a while. Complicating matters further—and this was where Jess's work came in—crevasses could crack to the bed and give the aquifer's water a path to get out.

And those were just the unknowns that we were aware of. Other processes could be happening that we hadn't discovered yet. "How well do we know what we don't know?" van den Broeke mused. "We cannot even answer that question with any certainty."

I'd already read many news articles about melting glaciers and ice sheets, of course, but the sense of impending catastrophe had always been tempered in my mind by an unwarranted optimism that humanity would figure out a solution. The more I talked to glaciologists, though, the more I started to genuinely panic. At one point, I wrote in an email to two friends, amid other life updates, that it was really alarming to realize how much we didn't know about sea-level rise. They didn't respond to that part of my message because, well, what could they say?

As my plane approached Kulusuk, mountains poked through the clouds, and brown rocks with white veins of snow rose from the sea like a slow-moving herd of animals. Icebergs were lit from below by a green, jewel-toned glow, as if I were seeing the northern lights underwater. We landed at the tiny Kulusuk airport: a one-room waiting area with a cylindrical metal door, Air Greenland and Icelandair desks in front of a salmon-pink wall, and a stand selling hot dogs, neon–red-and-blue drinks, HARIBO candies, and other sundries. Inside, it was surprisingly busy with what appeared to be a mix of locals and tourists with American and European accents. Through the large windows, the sky was so bright that it hurt my eyes to look at it directly; the drab landscape of rubbly tundra and tussocks of grass was punctuated only by the bright red suits of airport staff. Outside, the wind was bracing, and the dry air immediately turned my hair staticky.

From there, I boarded a helicopter to Tasiilaq that whirred over shimmering dark blue water, the mirror-like surface reflecting the rocks above. We soon approached rugged hills dotted with cheery red, blue, green, and yellow houses and landed at the town's heliport, a one-story building with a single gate labeled "Gate 1." The weather was relatively mild, and the landscape was brown with patches of green; even though Tasiilaq was only about a twenty-minute helicopter ride from the edge of Helheim Glacier, we were still a world away.

Jess would later tell me news she'd heard from a helicopter pilot—that the fjord in town had not completely frozen this year, as it usually did. Most of Tasiilaq's residents are Inuit people, who first settled on the island around 2500 BCE and migrated there in waves over the next few thousand years; they'd hunted along the coastline, and today their descendants continue using dog sleds. But people had started getting rid of their dogs, partly because they could no longer rely on ice conditions in the winter. The glaciers that people used to see from town were disappearing.

At around 7 p.m., I joined the scientists for a satellite phone call with their teammates who were on the ice. We met at the hotel, a sprawling building perched on a steep hill with a dining room overlooking the harbor. Jess, Colin, and the two guides, Mike and Richard, had flown to the glacier the day before; they'd asked Ian McDowell, a PhD student from the University of Nevada, Reno, to join them today to help with shoveling. Still in town were Winnie, PhD student Angelo Tarzona, and Jess's adviser Kristin, along with Kristin's husband and eight-month-old daughter. They'd assembled at a small table in an alcove with floral curtains; the baby was sleeping, and this was the only place they could meet within range of the monitor.

Kristin was calling the team on the ice, but no one was picking up. "You never like to know that your colleagues with a prearranged phone call time, who are very organized individuals, are missing their phone call time," she said. She sighed and went to the balcony to get a better signal.

"Still no call?" Angelo asked after she came back.

"No," Kristin said, sounding a little strained. Finally, a message arrived from Jess asking if they could talk at 7:30 p.m. "They're alive," Kristin said. "At least, one of them is alive."

Jess had told me earlier that Kristin would coordinate logistics in town while the other scientists worked on the glacier. She was relieved because, she said, Kristin was frank, direct, and had authority. "People listen to her," Jess said. I could see what she meant. Physically, Kristin wasn't imposing, but she exuded a cool, steady air and spoke in an unhurried, almost metronomic manner, like a clock that kept ticking at the same pace regardless of whatever rush and bustle might be happening nearby. She later told me that she got into glaciology when, as an undergraduate working in an astrophysics lab, she realized that she wasn't excited about the research. Rather than having an existential crisis, Kristin looked up science fields on Wikipedia and kept coming back to articles about glaciers and ice sheets. "That was how I objectively identified my interest," she said. Making hard decisions appeared to be her superpower.

The expedition had already run into obstacles. On their flight from Iceland to Greenland, approaching their destination with Helheim Glacier tantalizingly visible through the windows, the plane had turned around because it was too foggy to land at Kulusuk. Back at the airport, Jess injured her back while picking up a bag containing heavy equipment; after being stuck in Iceland for two days, the team's next flight was canceled, eating into their margin for error. After everyone finally made it to Kulusuk and then Tasiilaq—some team members scrambling for transportation on the last leg of the journey and riding a boat that rammed through the sea ice—the weather forecast for the glacier looked, in Richard's words, "quite unpleasant." Wind speeds were forecasted to be around thirty miles per hour and gusting into the forties, which would make the temperature feel like 0°F. When Jess, Colin, Mike, and Richard arrived at the camp site, snow blew constantly across the surface and pelted their faces with hard flakes; Jess felt like the wind was slapping her.

They spent hours digging and cutting blocks of snow to build walls to protect the tents. But during a late dinner in the kitchen tent, as they ate macaroni and cheese, drank some whiskey, and started to feel better, newly blown snow was quickly filling in one side of the outermost wall. Around 10 p.m., Colin went to the "bathroom," a pit in the snow, and saw that their sleeping tents were already getting covered. Everyone started shoveling; the wind had packed the snow into hard slabs, and they had to whack it and fling off heavy chunks. After already having spent the day shoveling to set up the snow walls, Colin was feeling a little delirious. To Jess, it looked like the snow was filling in so fast that her efforts weren't making any difference.

They'd finally gone to bed after midnight. Then at 3 a.m., Colin awoke to find that the tents were again getting covered in snow. Jess, in a restless sleep, heard Mike say, "I think we gotta wake Jess up." So she put on her cold, wet pants and shoveled for another hour. Mike said this wasn't sustainable. If the weather didn't improve, or they didn't change their camp setup, they couldn't stay.

In the hotel alcove, Winnie was checking the weather forecast. "This is bad, bad, bad, bad, bad," she said. She had a more restless energy than Kristin and described herself later as not having an intimidating personality; people at her university often thought she was a student rather than a professor. Originally from Hong Kong, she'd briefly entertained the thought of becoming an astronaut, but she abandoned the idea when her mom told her that wearing glasses would disqualify her. Instead, she'd studied geophysics in the UK, then joined a U.S. team that attached a "pod" containing a radar system and other instruments to planes in Antarctica. She liked radar, she said, because you could see through things. Before this trip, when I asked her if she ever felt like her research was too late to make a difference, she said no; it was more like "this might be the last chance." If she didn't act now, she said, "my grandchildren will not forgive me for just sitting back." As long as there was an ice sheet to study and scientists needed to answer urgent questions about how it affected sea-level rise, she'd keep doing this job.

"You'll be out on the ice sheet at age eighty-five?" I asked.

"I will definitely try," she said. "As long as I'm physically qualified to go, I will go."

When Jess got on the satellite phone, she sounded like she was in good spirits; the wind had died down. Winnie asked for an update on the two radar units that they'd installed during the last expedition. The instruments had been transmitting signals into the ice all year and recording the return signals; based on that information, her team could distinguish ice, snow, water, and rock and track the water's path through the glacier over time. But the units were buried under a few meters of snow that had fallen over the winter. The team needed to dig them out to retrieve the data and fix problems with the devices' satellite transmission.

"Winnie wants to know if you have been working on digging them out, and if so, how far you have gotten?" asked Kristin, who was closest to the phone.

"No," Jess said. "We've been busy all day." She said that if it wasn't possible to dig out last year's units, maybe they could set up a new radar unit instead and gather some data on the spot.

"No. No," Winnie said, standing up.

"If we can't get it out..."

"Can I talk directly?" Winnie moved toward the phone, sitting on the edge of the table with her arms crossed. "I'll tell you what our plan is," she said, her tone definitive—so that Jess could hear her clearly or that her point would be taken seriously, I didn't know. "The least amount of things I want to get done this year—"

"It's breaking up," Jess said.

"I need to dig two holes, one for the SISO, one for the MIMO, so we—one, get the data out, and second, try to fix the Iridium,"

Winnie said. "And that will be the minimum amount." SISO and MIMO were the acronyms for the two types of radar units, and the Iridium was the satellite transmission component.

Jess put Colin on the line to tell them about the snow conditions. "It's really hard," he said. "Basically concrete." Making matters worse, the team had run into a thick ice layer about five feet down while trying out a drill. "Digging a three-meter snow pit is, Mike says, impossible." Other team members at camp were dubious too, I later learned. Their backs were already sore from shoveling. And getting the radar units out wasn't just a matter of digging a narrow hole straight down; they'd have to shovel out steps toward the devices, like a terrace, to have a surface to sit or stand on and keep digging. An enormous volume of snow would need to be moved.

"I guess we'll have to figure out when we get there if it's even possible," Winnie said.

Kristin said, "Winnie, I think they're telling you that it's not possible."

Colin said the team was nervous about having Winnie and Angelo fly out. "With the storm approaching, we just want to make sure that they understand what they're walking into," he said. Winnie sighed.

"That is fair," Kristin said in a measured voice.

He suggested that Winnie come out for one day, on her own or with Angelo, and leave before the weather turned bad. "We'll consider that," Kristin said. Winnie got up and left the room. I started to think that if Kristin were the Jean-Luc Picard of the team—the calm, diplomatic leader—Winnie had a streak of Captain Kirk, who famously once said, "I don't like to lose." If she could have asked Scotty to beam her up to the ice, she would have.

After Winnie returned and Kristin hung up, there was a long pause. "All right," Kristin said, seeming like she was trying to find the right words. "It's not awesome, is it?"

Winnie got up and left again, looking upset, and then returned seeming calmer. What she worried about the most, she said, was if they didn't dig up a radar unit this year, it would be buried under twice as much snow next year. "And it will be forever lost," she said.

Kristin said, slowly and gently, like a kindly veterinarian telling a distraught pet owner that their beloved animal had to be put down, "I think it might already be forever lost."

No one spoke for a while.

Winnie said she wanted to ask another scientist, who had experience with the radar units, for suggestions. "Losing equipment is the last ditch," she said. "That you write it off so easily—you just don't."

"It sounds like you better go up there," Kristin said. "If I were you, I would take the one-day offer."

"I need some data back, as much as I can," Winnie said. She told me later that there was no way she could abandon the instruments without exhausting all options. "We might go out there and manage nothing," she said to Kristin. "But at least I've tried out everything."

When I met Kristin in the dining room the next morning, she told me that Winnie and Angelo were preparing their gear to fly out that day. Through the windows of the hotel, I could see the town below, mountains rising from wisps of clouds like a scene from *The Lord of the Rings*. (I'd brought a stubby paperback of *The Two*

Towers with me, and everything seemed tinted with a Tolkienesque light.) A layer of fog lay over the harbor in a strangely straight line, as if someone had cast a spell. Over the next week and a half, I would find that the character of the town changed from day to day. When the sun was out, the setting turned storybook-spectacular, the hills a deep green and an opalescent sheen on the water; at other times, everything was blanketed by fog, the buildings only indistinct forms. In certain lights, the low, angular icebergs in the harbor resembled sleeping white cranes with folded wings, as if a flock had flown over the mountain peaks and come to rest.

Later that morning, I walked to the heliport with Kristin's husband, who carried the baby on his back, outfitted in a pink knit coat with fuzzy ears. The town was laid out over a series of steep hills; a few streets met at a low crossroads over a river with rocky banks, which flowed from a valley with a cemetery to the west. We passed a soccer field and then headed away from the center of town, which had a post office, a church converted to a museum, an art workshop displaying traditional carvings, and a small grocery store where I'd later buy mini banana muesli bars imported from Denmark.

When we arrived at the heliport, a large patch of bare dirt with a hangar near the water, the team was loading gear into the helicopter: duffel bags, skis, solar panels. "This is good," Kristin said. "This is a good plan." Shortly before Winnie and Angelo were set to depart, Kristin saw a pair of forgotten gloves on the ground, grabbed them, and ran to give them to the pilot. Ominously, a raven croaked. But never mind—the helicopter was taking off, blades whirling and roaring, a red light flashing on the tail. Kristin jumped up and down. "Bye Winnie! Bye Angelo! Godspeed!" she called. The roaring became fainter until the helicopter was a white speck and disappeared into the sky as it headed inland, leaving behind the sound of twittering birds.

Arranged on a net on the ground were four of Jess's GPS units, big black containers to be carried on a sling load later. The pilots had already dropped off four units near their intended installation sites. For each station, Jess and Mike would drill holes in the ice, insert long poles, and set up a solar panel and antenna to receive GPS signals. Jess also planned to set up several units in a diamond shape extending away from the crevasse field; by tracking their movement, she could measure the stress in the ice that made it crack open.

Later, Kristin told me that she was worried they were wrong. Maybe water wasn't driving the crevasse they'd picked to the bed. She said, "What if we put these stations out, and at great expense, and time away from everyone's families, and sore backs from shoveling…" She meant, in essence, What if this was all for nothing? I didn't say it out loud, but as she was talking, I was thinking about a bigger fear: that this work would be for nothing not because their hypothesis was incorrect, but because humanity would pay no attention to their findings. That we would shrug, go about our lives, and change nothing.

When I'd talked to other glaciologists about whether improved sea-level rise projections would make a difference, some were more optimistic than others. Leeson speculated that policymakers' inertia might be partly due to the predictions' uncertainty; if scientists could narrow the range, "I think that they will be a lot more inclined to take action," she said. Other researchers wondered whether their time would be better spent helping figure out solutions for coastal adaptation.

At one point, I'd asked Joanna Huxster, a researcher at Eckerd College in St. Petersburg, Florida, if the glaciologists' efforts were a waste of time. Huxster studies public understanding of ideologically divisive science, including environmental issues. When she and others in her field finally convince people to take action on climate change, she said, "I want those accurate sea-level rise projections to be ready." Sometimes people asked her, "Why are you hopeful?" And she answered, "What other choice do I have?"

After the previous night's tension, the problem of the buried radar units was resolved with surprising speed. Winnie and Angelo arrived safely at camp and the team immediately started digging. Once they reached the ice layer, they sawed through it with a power tool and hammered it into pieces. After they'd flung out chunks of ice, their task suddenly became much easier. Underneath was sugar snow, soft as pudding. About eight feet down, Ian poked a probe into the snow and heard a *thoonk*. He'd hit a plastic Home Depot storage container lid placed on top of the box containing the radar unit. When Colin pulled the device to the surface and Winnie plugged the SD cards into a laptop, she saw that the instrument had collected data for nine months, from September 13 at 1:43 p.m. to the moment they'd retrieved it, June 11 at 5 p.m. It had worked.

When Kristin heard the update from Jess at the team's evening call, she kept saying, "Wow. Wow." After they hung up, she said happily, "What could be better?" She couldn't believe they'd gotten the radar unit out. "I'm tickled pink."

The next day, I boarded the helicopter for a quick "boomerang" flight to camp to pick up Winnie and Angelo and bring them back to town before the next storm arrived. At the heliport, Kristin had packed a funnel, a pair of gloves, a bottle of Clément French Caribbean rum, and a pack of cards for me to deliver. The pilot gave me safety instructions that I definitely wasn't going to remember, including which button to press to send a rescue beacon. "Just in case I'm not…" he said, presumably implying "conscious." The rotor blades whirred, the helicopter shook, the floor vibrated, and we were off.

We raced over the water toward the glacier, sun lighting up the cabin. The helicopter was headed to latitude 66°21′ north, longitude 39°9′ west, a magenta line marking our route on a map on the pilot's tablet. The town fell behind us and we flew between cloud-topped mountains; brown rocks gave way to a stippled bluish surface. At the glacier's edge, the ice thinned into a delicate sheet over the water. Then we were over the ice sheet, all color erased and the landscape leveled by the sheer force of its existence. The only other features were pale blue sky, light streams of clouds, blue-gray mountains ahead. Below, the surface was faintly etched as if with a paintbrush; then deeper gouges appeared, like the tracks of an animal dragging its claws. At the bottom of the glacier, water was likely flowing in a thin film over the bedrock, meandering through pockets on the ground, and converging into riverlike channels.

The ice in Greenland seemed to have infinite variations, like light. On the flight to Kulusuk, the soft, dreamy swirls of sea ice fragments had resembled cherry blossom petals scattered over a meadow. On a boat ride between two islands, I'd later pass jutting icebergs up close; two rose from the water like sentinel towers, all geometric shapes and dramatic angles. Now, on the glacier, everything flattened into a never-ending sweep of white. It was as if three wildly different artists—one Impressionist, one Cubist, and one Rothkoesque—had been asked to paint what they thought ice looked like.

After about fifteen minutes, we descended toward camp, the tents at first a smudge in the distance and then two red domes coming sharply into view. On the ground, the camp had an air of busy worker bees. Snow-bricked walls protected one side; the kitchen tent was stocked with a stove and plastic bins of food, and a communal sleeping tent was scattered with rumpled sleeping bags. Every evening, Richard set up a nylon trip wire encircling the tents, rigged with cartridges that would detonate if a polar bear tried to cross the perimeter.

There was little time to look around, though; we had to leave camp almost immediately, before Tasiilaq got fogged in. Jess and Mike were working at a GPS site between camp and the crevasse field, and Richard wanted us to drop off food for them. After he handed Winnie a peanut butter, jam, and cheese sandwich in a plastic baggie, she and Angelo and I piled into the helicopter and headed toward the GPS site, where I saw two small figures on the ice. The pilot briefly touched down on the ground and tossed the sandwich out the window, the wind blowing it like a tumbleweed toward Jess and Mike. They waved. I couldn't see Jess's face, but she seemed happy.

As we headed back to town, we passed the crevasse that Jess planned to instrument. It came into view as the pilot swung around, a gaping blue maw that looked big enough to drive a semi-truck through. The GPS containers were on both sides, from our vantage point appearing to be little more than black

LEGO bricks. I couldn't see the end of the crevasse; it looked like the underground lair of a giant ice monster that crawled through tunnels in the glacier, drinking from the hidden aquifer and melting a path to the surface with the heat of its breath.

I thought about what Jess and her team were trying to do—to understand the ice—and it reminded me of my process as a writer. I grasped for metaphors to describe the glacier, took dozens of photos, and scribbled phrases in my notebook, in an attempt to convey perfectly what I saw. But I'd reached the limits of my capabilities; I'd never be able to capture it as it truly was, in all its beauty, stillness, and power. And the scientists would reach that limit too. To try to represent the ice sheet in their equations and computer models seemed like an act of hubris—or maybe it was hope. I decided to go with hope.

Jess never got to the crevasse field during her remaining days at camp. Soon after I left, rain and blowing sleet started. With the precipitation, clouds, and fog, the scientists couldn't see the horizon; they had no reference points to judge direction, no trees or rocks or incline in the terrain. Skiing to the crevasses was too dangerous. One day, as the rain fell, Jess and Mike spent several hours installing two GPS stations in the diamond closer to camp. By the time they were setting up the second unit, Jess's clothes and gloves were soaked, her hands freezing.

"Everyone is pretty much wet to the bone," Colin said to Kristin at the morning check-in call the next day. "It's basically like taking a shower, it's just colder." They had to minimize their time outside; if people ran out of dry clothes, they could get hypothermia. And they were still in whiteout conditions.

"We're basically just scrapping the crevasse GPS?" Kristin asked.

Colin said that after the team returned to town, Jess and Mike might fly back to the crevasse field in a few days, on Monday, if the weather was good. But for now, the plan was to finish installing the weather station and a device to measure snow temperature at various depths, then "get out of there." They'd gotten some good science done; the team had gathered more seismoelectric data, taken density and permeability measurements, dug out the second radar unit.

I was worried that the crevasse GPS units kept getting put off, but Kristin seemed unfazed. "The Monday plan is a good one," she said. They'd gone from zero units last year to installing part of the diamond this year, which was "way better than nothing." On Friday, the sky cleared by the afternoon and the team made it back to town. When I arrived at the heliport, they were unloading and drying out gear; Mike handed around green cans of Danish Tuborg beer. "Cheers!"

The morning dawned on Monday—the day Jess would finally set up the crevasse GPS units—with beautiful sunny weather. At the heliport, I asked her how she was feeling. "Good," she said. "Excited." She, Mike, and Colin unpacked equipment from the shipping container, donned safety gear, went through their checklist of items. By the helicopter, Mike tightened Jess's and Colin's harnesses like a groomsman straightening cuffs before a wedding. There was an air of expectant waiting. No one was saying anything.

"Don't fall in," I said.

"Hopefully we do fall in!" Jess said. She wanted to rappel into the crevasse to see or hear water flowing. Several minutes later, the pilot appeared. "All right, shall we go?" he said.

"Yeah!" Jess said.

The team piled in and the rotor blades started whipping around, that air of expectant waiting again, the moment before the moment—it seemed to go on forever. The sound deepened, and dust blew up from the ground. Suddenly the helicopter took off, looped around once, and wheeled toward the mountains. The plan was to return in the late afternoon. But at the hotel bar around 1 p.m., while I was working on my laptop and Kristin and her husband were feeding beets and pickles to the baby, Kristin got a call from Jess's team. They were coming back early. Fog was on its way, which might prevent them from flying back to Tasiilaq, and they'd heard that the Kulusuk airport—their backup place to land—might close too.

After the team returned, I gathered that they'd managed to install one GPS station. They had to move two other units that were dangerously close to a second crevasse but then ran out of time to set them up. Jess told me later that they'd been so close; another couple hours on the ice sheet and they would have finished. But Mike had other commitments he needed to return to the States for, and everyone was flying back to Iceland the next day. "So we got six installed?" Kristin said, meaning the total number of GPS units. "That's pretty good." She asked whether they'd set up unit number one, two, or three.

"One," Jess said. "The tip of our diamond."

"The one on the far side? Nice. Good decisions." At least they'd completed the diamond; but to monitor whether the crevasse was widening, they still needed one more unit. The pilot, after swapping stories with Jess about research obstacles—bears attacking a weather station, chewing the rubber off cables—looked sympathetic. "It's hard to do fieldwork, huh?" he said.

The following day, when everyone was wrangling with more logistical hurdles to catch flights home from Kulusuk—fog, helicopters delayed or canceled, trying to take a boat instead, but no guarantees it could get through the sea ice, and so on—Colin kept repeating, "Expectations lead to pain." I suggested to Jess that this would be a good motto for fieldwork. Another scientist, who'd been studying a different part of the glacier, said, "You have to have *some* expectations." Jess said, "But there will be pain."

A month later, Jess emailed me with surprising news. Leigh Stearns, a glaciologist then at the University of Kansas in Lawrence, had traveled to Helheim after Jess's team left. As a favor, her team had set up the last two GPS crevasse stations. I couldn't believe it; Jess had gotten her happy ending. And she would return to Greenland by herself in September to download data from the units, the helicopter ferrying her from station to station.

Of course, Helheim had other plans.

When she reached their campsite, Jess found the GPS station tilted, the black container flipped over and buried under the snow. Other stations between camp and the crevasse field were in similar disarray: the solar panel face-down, the antenna not even visible. That summer in Greenland had been unusually warm; instead of six inches of snow melting, as climate models predicted, roughly six *feet* had melted at their field site, making the surface too unstable to hold up the units. Conditions were changing so fast that it was getting harder to do fieldwork, Jess said: "We're just seeing things that we've never seen before."

There was no time to repair the units and set them up again, so Jess salvaged whatever data she could and the pilot flew her to the crevasse field. The surface, mostly smooth in June, was now scored with jagged openings. Circling over the first GPS unit, they realized Jess wouldn't be able to work in the area safely without a rope team; a new crevasse had opened up, and the instrument was perched on a snowbridge over it. And there was no safe place to land the helicopter near the second and third units.

But Jess said there was still positive news. These GPS stations hadn't fallen over. And the crevasse that Jess had chosen to instrument had widened from about thirty-seven to forty-three feet—exactly what she'd been hoping to capture. Plus, more crevasses had opened between the second and third units. When she came back next summer with more people, they could form a rope team and ski safely around the cracks. Once they retrieved the instruments, "we should get some really great data off of them!" she wrote in an email. "So I am very excited about that."

On a call with Jess, I brought up the issue I'd discussed with other glaciologists—that accurate sea-level rise projections wouldn't help if policymakers did nothing. Jess was always so upbeat that I thought she'd say something optimistic. But when I asked if she ever questioned whether her work was worthwhile, she said, "I do." She'd gotten into glaciology a decade ago, wanting to make a difference. People in power still weren't doing much, and we were seeing one climate disaster after another. "What can we do to make them listen?" she said. "I don't know if we can do anything." The sad part was that scientists coming together to study climate change "hasn't seemed to be enough," she said. "It's not enough."

I thought about Jess spending months preparing for the expeditions, the constant delays, she and Aleah huddled in their tents with COVID, the team hacking at snow at three in the morning, digging eight feet below the surface to get the radar data, Mike and Jess setting up GPS stations in relentless rain, Jess drenched to the bone. But that still wasn't enough.

Still, as long as scientists were trying, I had to believe that their work would mean something someday. Even if it was just a historian hundreds of years in the future, surveying the remnants of a weakened civilization beset by climate refugees and battles over food and land, sifting through hard-won scraps of data gathered in the hope that science would save the world—that wasn't too dramatic a phrase—maybe that person would say, *Well, they tried. They really did try.*

Or perhaps that person was living in a society that had managed, just barely, to pull back from the brink. Maybe humans had awakened to the crisis, listened to scientists, paid attention to those new sea-level rise predictions, used the knowledge to save themselves. Maybe all of that had, improbably, come to pass. And maybe my imaginary historian was reading about what Jess and her team had done hundreds of years earlier, and still said, *They really did try*, but also, *Thank you.*

Fig 1. Male and female vocalizations

Evolutionary Biology

Call of the Guardian Frog

"It sounds stupid, but I really like watching animals."

In 1984, a young man in a tropical rainforest came across a small frog. He was wading in a stream in Borneo, a massive island teeming with life between mainland Asia and Australia. Perched on a rock that jutted over the water, the inch-long creature had pebbly brownish-black skin and shining, gold-rimmed eyes. It was a male smooth guardian frog, a species that had been briefly described by a biologist about a century earlier but easily escaped notice by blending into dead leaves on the forest floor.

The young man, one of the island's Indigenous Iban people, was working with visiting scientists from a natural history museum in the United States. He collected the frog in a plastic bag and noticed something curious: Half a dozen tadpoles, presumably the frog's young, had ended up in the bag as well. When he showed his discovery to the researchers, they thought that the adult frog had likely been carrying the tadpoles on his back and the little creatures had fallen off.

Two years later, scientists from the museum returned to survey species with local workers in another part of Borneo, and the team discovered two male smooth guardian frogs hiding under leaves in the forest. Each frog was tending a clutch of eggs. They also captured three males with tadpoles clinging to their backs near streams, confirming their earlier suspicions. The researchers surmised that male smooth guardian frogs took care of the eggs, guarded them from predators, and, after their young had hatched, ferried the tadpoles to water.

They wrote up their observations in two short reports for a specialist scientific journal. Then for the next few decades, little was published about the species' behavior. The smooth guardian frog was largely forgotten.

In 2011, Johana Goyes Vallejos read the two reports. Johana was a PhD student in ecology and evolutionary biology at the University of Connecticut in Storrs; originally from Colombia, she had grown up in a city in the Andes mountains and had never been to Borneo. In fact, if you were to drill a hole through the earth from her hometown to the farthest possible place on the other side of the planet, you would end up right around Southeast Asia.

Johana had heard about this frog from a German scientist she'd met at a conference, who lived in Borneo. He found it intriguing that the smooth guardian frog was one of only two frog species in the region known at the time to care for their young. Many scientists had observed poison frogs tending to eggs and carrying tadpoles in Central and South America, including colorful species with neon green, red, blue, and yellow skins popping like superhero capes against the forest landscape. But this brown, nondescript frog in Borneo piqued Johana's interest. She wanted to study the species that no one else wanted to study.

As it turned out, there was a good reason that no one else was studying it.

For one thing, smooth guardian frogs are masters of camouflage. Unlike the brightly colored poison frogs, which are toxic to predators and hop about boldly, these creatures are vulnerable, shy, and secretive. The frogs are nocturnal, so Johana would have to search for them at night. And they live over a large swath of rainforest; she couldn't just go to a pond to find them. If you listed all the features that would make an animal's behavior ideal to study—numerous, easy to locate, and so on—the smooth guardian frog checked none of those boxes.

But Johana was curious to learn more, not just about the tadpole-carrying behavior but also something else the German scientist had mentioned. One of his students had heard a female smooth guardian frog responding to a male's call with her own call. This was unusual; in most other frogs, the females were generally thought to remain silent, except to protest when an unwanted male grabbed them. Among the roughly six thousand species of frogs and toads known at the time, scientists had reported only a couple dozen examples of females vocalizing during mating, usually replying to males to reveal their location. Johana wondered, Why were these female frogs calling? She decided to fly to Borneo to investigate.

During fieldwork trips over the next few years, she found that smooth guardian frogs were far more interesting than she had even thought initially—that they might represent a rare case of a phenomenon called sex-role reversal. In sex-role reversed species, females compete intensely for male mates, often because males are occupied with caring for young. Scientists had seen this set of behaviors in only a tiny fraction of the animal kingdom and had never found a definite case in frogs.

Once she had started gathering bits and pieces of information about the frogs, Johana became consumed with trying to understand them. In graduate school, each time she traveled to Borneo, she moved out of her apartment and stored belongings in her office to save money on rent and scrounge up the cash for field station fees. She had almost no funds to buy equipment; a Sony Handycam to film the frogs, purchased by her adviser, became her most precious possession.

I first learned about Johana's research in 2018, shortly after she'd published new findings describing how the male frogs carefully tended to their clutches of eggs. A few years later, I got back in touch, and she told me that she still had so many questions about the frogs' behavior. But she'd been stuck, first unable to find a research job after finishing her PhD and then thwarted by logistical delays, and she hadn't visited the forest for years. Johana worried that the frogs would vanish before she could learn everything she wanted about them. "I don't even know if they are alive," she said. "For all I know, they could have gone extinct." With climate change and other threats looming, she felt like she was running against the clock.

After a failed attempt to travel to Borneo in 2022, she returned home with nothing but was still determined to get back to the rainforest. She said, "The day I hear these frogs calling, I will cry."

Frogs have a reputation for being laissez-faire parents. In North America, for instance, a typical frog "lays a million eggs and is like, 'Bye, good luck,'" said Eva Fischer, a scientist who studies the neurobiology of frog parental behavior at the University of California, Davis. But some frog species don't simply abandon their offspring—around 10 to 25 percent, depending on the estimate.

Scientists started noticing this behavior more than three centuries ago. In 1705, German naturalist and illustrator Maria Sibylla Merian published the first known observation of parental care in frogs. In a rare case of a woman undertaking a scientific voyage on her own, not as a man's companion, Merian had traveled with her daughter to Suriname in South America. There, she found *Pipa pipa*, an aquatic frog species with a flattened body. In what one scientist later described as "a spectacle that is the stuff of nature documentaries and horror films," the mother's fertilized eggs were embedded in pockets in her back, grew into froglets, and emerged from her skin.

Other early reports were equally bizarre. In the late 1700s, a French doctor observed a male toad with fertilized eggs wrapped around his legs, carting around his offspring like a string of pearls. "Nobody believed him," said Lisa Schulte, a biologist who studies behavior and evolution at Goethe University Frankfurt in Germany. At the time, scientists generally assumed that frogs weren't capable of taking care of their young: "The thinking was that amphibians are a stupid creature that wouldn't do that."

But over the next couple centuries, biologists continued to uncover a dizzying number of parenting strategies. Some frogs dug a little burrow in the mud for their eggs, and then excavated a channel to a nearby water body where the tadpoles could finish growing. Others whipped up foam "nests," like egg whites, from a substance that the female released; these frothy enclosures protected eggs from predators, microbes, and ultraviolet light. Some frogs guarded eggs from hungry insects or spiders, or they released water from their bodies onto the eggs—essentially peeing on them—to keep them moist. And in one species in Chile, the male swallowed the eggs and let them grow inside his vocal sac, after which the froglets exited their father's mouth.

In fact, among frogs that looked after their young, males often turned out to be the sole caregivers. When biologists examined data on 276 species with parental care, they found that only the father looks after eggs or tadpoles in more than half of cases, compared to about one-third with female-only care and one-tenth with both the male and female participating. Similarly, in fish, males are much more likely to take on parental duties.

Humans tend to assume that female parental care is the norm because that's what happens in mammals. But the research in frogs and fish "tells you that it's not something that females are more prone to do," said James Tumulty, a behavioral ecologist at Rhodes College in Memphis, Tennessee, who has collaborated with Johana. "There's not something about being a female that necessarily predisposes you to providing parental care."

As the mom of a then nine-year-old, I was dimly aware that being impressed by male frogs caring for their eggs was the equivalent of a well-meaning onlooker telling a man on a routine errand with his kids at the grocery store, "You're such a good dad!" Still, I couldn't help but find something refreshing about behavior in other animals that flaunted our society's default roles. The feeling was similar to the pleasant surprise I'd felt when reading a science-fiction novel describing an alien culture, in which males and "shons" (individuals who switched between genders) usually assumed the parenting duties; these caregivers were professionally trained in skills such as first aid, math tutoring, and storytelling and worked full-time on raising the species' young after females handed over their offspring.

None of the findings about frog parental care would have emerged without simple observations of animals: just sitting there and watching what they do. But Schulte noted that this kind of science was falling out of fashion; research testing a hypothesis was considered more important. These days, she said, it was harder to publish studies that only said, "Ah, I think I saw a weird behavior in this frog."

For Johana, her passion for frog research traced back to that basic desire to observe. She had grown up in a highly protective family, not allowed to play on the monkey bars or the merry-go-round, often dressed in pink dresses with puffy sleeves and shiny patent leather shoes. And she didn't see much wildlife in her city. But Johana had read about pandas and opossums in her aunt's *National Geographic* books, played with clams at the beach, touched a dolphin's rubbery skin at an aquarium. She decided to study biology in college and was desperate to join a fieldwork expedition, to break out of her hermetic bubble. During one class field trip, a couple of professors asked what she wanted to do for her career. "I really like watching animals," she said. "It sounds stupid, but I really like watching animals."

When Johana arrived in Borneo on her first trip in 2012, the forest was like nothing else she'd ever seen. She'd gotten her wish to do fieldwork after her undergrad program, counting sloths on an island off the Pacific coast of Colombia and recording frog calls in a Panama marsh. But this was a different experience entirely.

To reach the rainforest, she'd gotten a ride from Connecticut to New York and flown to Frankfurt, Germany, then Singapore. Borneo is part of an archipelago that includes Sumatra, Java, Bali, and the Philippines; the world's third-largest island, it almost seems like a small continent. Johana was headed to Brunei, a small country on Borneo's northern coast; the rest of the island was divided between Malaysia and Indonesia. From Bandar Seri Begawan, the capital of Brunei, she rode a speedboat through mangroves and then traveled by van to a pickup spot where another boat transported her along a river toward the heart of Borneo. After about an hour and a half, she arrived at the Kuala Belalong Field Studies Centre, a set of buildings on the riverbank connected by wooden boardwalks. Set in the midst of a national park, the station was surrounded by steep, rugged slopes, all blanketed in thick greenery.

The trees in the rainforest rose so high that Johana's neck hurt when she looked up at them. Borneo is home to the world's tallest tropical trees, with some soaring to more than 300 feet. Many animals had evolved to traverse the towering forest by gliding from tree to tree: Wallace's flying frogs, named after the naturalist Alfred Russel Wallace, which used their webbed feet like parachutes; snakes that flattened their bodies into ribbons and "swam" through the air; small dragon-like lizards that spread their ribs and flared out flaps of skin so they could coast like Frisbees.

Swooping, scurrying, and slithering among the trees were birds with curved, reddish-gold "horns" protruding above their beaks, petite nocturnal primates with huge dinner-plate eyes, beetles with feelers resembling long bobby pins that they fanned out and folded away, emerald-green vipers. The forest was occasionally punctuated with the sounds of arboreal gibbons, *wwwwwooop wwwwwooop wwwwwooop whoo-ooo-ooo*, and the barking of geckos, *ruk-roooowr*. Around 6 to 7 p.m. every night, chaos erupted with the loud calls of insects and the continuous whine of mole crickets—*neeeeehhhh*—amplified by the holes they lived in. The forest was warm and humid, filled with the scents of wet soil, rain, rotting figs, and musty decomposing leaves, with occasional wafts of ginger. When darkness fell, clusters of mushrooms glowed a luminous green.

But traipsing through the forest required care. At night, wasps that were famous for excruciating stings became aggressive when they saw lights. During one of Johana's later fieldwork expeditions, when she was hacking a trail with a machete during the day, she accidentally disturbed a group of the insects and was immediately stung half a dozen times in her arms and over her left eye, which swelled up so much that it looked like someone had punched her. The biggest risk was the trees: They were so tall, and their roots so shallow, that after a heavy rainfall, a falling branch or tree could kill you. Once, Johana returned to a tent in the forest where she'd been working the previous night and found it had been crushed by a giant branch. If she'd been inside at the time, she could have died. Another time, she looked up from the ground and found a dark green, coiled viper staring her in the face. Still, she figured she couldn't be thinking all the time about the risks; she could die after tripping on a bar of soap in the shower too.

Johana had traveled to the field station that first year with Ulmar Grafe, a behavioral ecologist at Universiti Brunei Darussalam in Bandar Seri Begawan, the German scientist who had originally told her about smooth guardian frogs. Along with Hanyrol Ahmad Sah, a PhD student in Ulmar's lab, they started searching for the frogs in the forest. Johana wanted to observe the males' calls in more detail and figure out why the females responded—for instance, if they were telling the males their location.

But the frogs weren't calling. Night after night, the scientists trekked through the forest from about 5 p.m. to midnight and heard nothing. Johana started to panic: She didn't even know how to find the frogs, and this endeavor was going to fail miserably.

Finally, after what felt like an eternity, they heard a distinctive noise one night. It was a rapid, insistent, long trill, like a duck quacking repeatedly but sped up to sixteenth-note speed; Hanyrol later said it sounded like a phone ringtone. The call was so loud that they could hear it from far away. "That's it," Ulmar said. They ventured off the forest trail, turned off their lights, and crept toward the trill. When they reached the source of the call, they turned on a light and saw a male smooth guardian frog, the sound coming from a creature tiny enough to sit on Johana's thumb.

Over the next few weeks, Johana and Hanyrol gradually found more smooth guardian frogs and observed their behavior. When the male frog was calling, he inhaled gulps of air to inflate his body, then released the sound with much puffing and trembling of the skin under his mouth, his entire body vibrating mightily with the effort. The researchers' progress remained torturously slow. Even if they heard a call, the frog could easily leap hundreds of feet away in a few minutes, and chasing it was futile. Sometimes, after searching for seven hours, Johana returned to the field station defeated and then heard a male's trill as she was lying in bed. That was the worst; she knew she probably wouldn't find him at this point if she got up, and she couldn't bear to hear him and know that he was beyond her reach.

But the female frogs captured Johana's attention the most. One night, after finding another male, they had heard short, soft, high-pitched *bleep*s, like the sound of a child pushing on a squeaky toy. The noise was coming from a nearby female smooth guardian frog, her body twitching a bit with each call as if she had the hiccups. She wasn't just responding to the male this time, as Ulmar's student had reported earlier; she was calling on her own. Then the male replied with a new sound—instead of the loud trill, a brief, tentative-sounding *quack*.

A few weeks later, the rain was pouring so hard that Johana couldn't go out to search, so she was watching a movie in the field station lab. Her captured frogs were in glass terraria under the lab bench. At around 11 p.m., the females started *bleep*ing, and a male *quack*ed back. But what surprised Johana was that the females kept calling over, and over, and over: *Bleep, bleep, bleep. Bleep, bleep, bleep.*

To hear the female of an animal species calling repeatedly may not sound like a big deal. But among frogs, "most of the time, the females don't call at all," Tumulty said. "I've never observed or seen a female frog call in my career." ("*Never*?" I said incredulously. "No, never," he said, at least not in the species he'd studied intensively.) And finding females that vocalized spontaneously, not as a response, "has hardly ever been reported," said Johana's adviser Kentwood Wells, when I spoke to him in 2018; Wells, who passed away in 2024, was a herpetologist at the University of Connecticut and the author of a 1,100-page tome about amphibian behavior and ecology. "That's so rare in frogs." The female smooth guardian frogs also called much more frequently than males did—something that, as far as Johana knew, scientists had never seen in frogs before. Forgive the anthropomorphism, but if you imagine that most women around the world hardly ever spoke except to ward off an unwanted advance ("get lost!") or to reply when men called them ("yes, dear," "I'm over here, dear"), while the men around them chatted volubly; and that one day scientists came across a small, remote village where all the women talked on their own, and in fact spoke up far more often than men did—well, that would give you some idea of how surprising these findings were.

During that field season, and a second trip the following year, it became clear that female smooth guardian frogs were unique in other ways too. In most frog species, when the males call for mates, they try to outdo one another to call the loudest, most frequently, and over the longest period of time; a female then chooses and approaches a male. And sometimes, males grapple with each other to get to a female. Early research on animal behavior suggested that this pattern was nearly universal among not only frogs but other creatures as well. Males compete, often by fighting or developing showy "ornaments"—for instance, birds-of-paradise's brilliantly hued plumage—and the female selects her partner or accepts the winner of the fight. In other words, *The Bachelorette*, not *The Bachelor*.

But when Johana and Hanyrol set up speakers in the forest playing the sound of a male call, multiple females gathered around the speaker. And sometimes Johana found one male frog surrounded by several calling females.

It was at this point that she started to wonder if she was seeing a case of sex-role reversal. Scientists had observed the phenomenon in certain species of fish, birds, and insects. But the number of confirmed cases in frogs was precisely zero.

What exactly is sex-role reversal? The answer, it turns out, is not straightforward. When I asked biologists to explain it, my eyes quickly glazed over with their talk of abstract concepts like the operational sex ratio and something called the Bateman gradient, but the idea became easier to understand when we started chatting about pregnant fish and cricket sperm packets.

First, consider a group of fish called syngnathids, which includes seahorses and pipefishes. In some species, the males carry eggs in a special pouch and exchange nutrients and oxygen with the embryos. This gestation process is essentially a pregnancy. Meanwhile, the female doesn't help with any parental duties and usually just takes off. "All the female has to do is cruise through, give a male a bunch of eggs, and then she can go off and live her life and never have to worry about those offspring again," said Robin Hare, a lecturer at the University of Western Australia in Perth who has studied sex-role reversal.

Second, consider a species of Australian bushcricket, *Kawanaphila nartee*, which Hare described as a "small, quite sweet, dopey-looking" insect. The male cricket produces a "nuptial gift" to give a female: a sperm packet surrounded by a big, sticky mass of proteins that weighs about 20 percent of his body mass. During mating, the male plugs the nutritious present into the female's rear end, and she curls around in a back-bending gymnastic maneuver to eat it while the sperm transfer into her body.

What do the seahorse and the bushcricket have in common? In both cases, the male invests a lot of time or energy in reproduction, either by caring for young or coming up with a nuptial gift for his mate. (In other species, such as dance flies, the gift is a prey item that the male catches.) These activities keep the males busy for a while, so fewer males are available to mate at any given moment.

As a result, females are left scrambling for the remaining males. And faced with this supply-and-demand issue, females must compete with one another. For instance, in one pipefish species, the females boast a flashy, colorful skin flap that they unfurl during courtship to attract mates. In certain dance flies, the females inflate sacs on their abdomens with air to make themselves look bigger and gather in groups to show off to males. The mating roles seen in most early animal behavior studies get flipped: Females compete for males more intensely than the other way around.

The number of animals with strong evidence for sex-role reversal is vanishingly small—perhaps a few dozen among roughly 1.6 million cataloged species. That's less than one-hundredth of 1 percent, "even if scientists have massively underestimated the true number," Hare said. "It's minuscule."

And scientists haven't found any frog species with definite

sex-role reversal yet, in spite of the fact that males often take care of young, because the mating roles don't completely switch. For instance, in the green and black poison frog, males take care of egg clutches, and females act aggressively—but the males keep competing too. They call for females and mate with multiple females while looking after the eggs, spreading their attention among several clutches. "They do a lousy job," said Kyle Summers, an evolutionary biologist at East Carolina University in Greenville, North Carolina.

So if Johana could show that the smooth guardian frog was the first frog ever identified with sex-role reversal, I wondered, what would that mean for science? Alas, the complications did not end there. The further I waded with biologists into the topic, the more I sank into a quicksand of questions about our cultural biases and assumptions: Should sex-role reversal even be considered a thing? Should male and female sex roles be a thing? Why are we so obsessed with the differences between males and females anyway?

To begin with, the entire concept of sex-role reversal obviously requires that a species' behavior be "reversed" from "typical" behavior. But biologists have started to question the long-standing view that the vast majority of animals follow the "males compete, females choose" mating system. "I feel like we just don't actually know that," said Sara Lipshutz, a behavioral ecologist at Duke University in Durham, North Carolina. Past research was heavily influenced by Charles Darwin's writings; he described females as "coy" and "passive," while males were "eager" and had "stronger passions." While he did point out exceptions, his general framework—competitive males and choosy females—steered generations of biologists. "We're just not asking how widespread these other behaviors are," Lipshutz said. "And when we do search for them, we find them." Studies over the last few decades suggest that animal behavior lies along a continuum, with females sometimes competing and males sometimes choosing, rather than at two extremes.

Previous research on animal behavior also tended to be skewed by cultural baggage and scientists' viewpoints. For instance, biologists thought for a long time that pairs of birds that parented together were "paragons of monogamy, it's so beautiful," Fischer said. Then genetic research revealed that females were often mating with multiple males. Since female birds resisted their new partners in some species and were thought to be passive in mating, scientists assumed that "extra" couplings usually took place against the female's will. But in the 1980s, biologist Susan Smith reported that female black-capped chickadees were the ones initiating their affairs. Rather than simply succumbing to new males' advances, females snuck out of their territories to pursue higher status mates. "I don't think that many female biologists would question the agency of a female animal," Lipshutz said. "But historically, women weren't doing science."

Bias can go in the other direction as well: The language that scientists use to describe animals' behavior can influence what we expect of people. The term "sex roles" carries cultural connotations, said Malin Ah-King, an evolutionary biologist and gender researcher at Örebro University in Sweden. The risk is that scientific phrases could be used to define "natural" roles in society—for instance, to argue that aggressive male behavior and passive female behavior is typical, and "reversed" behavior is not. "Science and nature and animals are often used to naturalize what is normal and what is not normal," she said.

And some researchers think that scientists should move beyond slotting animal behavior into two sex roles. Among humans, some purported differences between the male and female sex—for instance, that boys are naturally better at math—have already been contested or contradicted by evidence. "And yet we haven't really interrogated that in nonhuman animals," Lipshutz said. If scientists always look for differences rather than similarities, "we're just reinforcing this expectation that we already have."

Still, studying species that do show rarely reported mating or parenting behavior could help biologists understand the rich diversity of the animal kingdom. In the past, when scientists' observations didn't match their assumptions, they would sometimes come up with convoluted explanations to "make it fit," Fischer said. By paying more attention to these differences and when they emerge, "that's how we really begin to understand how evolution works."

So perhaps the smooth guardian frog was, in fact, a rarity among Earth's creatures. Or perhaps no one had looked hard enough for this behavior before. Perhaps Johana was simply the first to notice.

On her third expedition to Borneo in 2014, Johana was eager to observe the male frogs' parenting more closely. She got nine captured pairs of frogs to mate in terraria at the field station lab and then waited to see what happened. Two clutches of eggs were left unfertilized, another looked sickly, and one male escaped; but males sat still with the remaining five clutches for about a week and a half. When Johana left cameras running to record their behavior, sometimes she thought they were broken because nothing was moving. The males ate only their own shed skin or the occasional insect or spider that walked in front of them. And they never called for another mate, as the male frogs in other species sometimes did during parental duties. "They are very good fathers," she told me.

When their young were ready to hatch, the males started rolling and tapping the eggs, making the tadpoles spin inside. Then they stepped on the eggs to break them open. "It's kind of like waking up the children to go to school," Johana said. "Everybody get out, we need to go!" The tadpoles, which looked like little drops of gelatin with tails, wriggled onto their father's back and arranged themselves in an organized formation. Johana wasn't sure how they stayed on, but the wet tadpoles may stick to the father's skin through adhesion. "It looks like they just hold on for dear life," she said.

Sometimes, she spotted males carrying their tadpoles in the wild. Since she had previously identified some of them by clipping their toes, she could tell whether she'd seen them before. "I always loved finding them," Johana said. "It was cool to know one of my frogs had become a father."

In one picture that she took, a male frog crouched under a ceiling of brown leaves, his skin glistening with moisture. About a dozen shiny, gelatinous egg capsules were arrayed in front of him, with grayish tadpole shapes forming inside. The male's protruding eyes—large black pupils circled in gold—seemed too big for his head, which I supposed was normal for frogs but gave him an air of deep watchfulness.

In a video of a male attending to his eggs, he started with the egg clutch positioned protectively under his chin, and then slowly inched closer to cover them with his body, legs splaying to make a little tent. Once in a while, as pinprick pairs of black eyes became visible inside the eggs, he shifted and rotated, moving so that the eggs were under his chin again, his body quaking a little. Who knew what was going on in that frog's brain, whether he was feeling something akin to fatherly emotion or simply operating on base instinct, nothing like the complicated tangle of excitement, anticipation, nerves, and fear that humans experienced? We'd never know. And yet watching the video, seeing him turn and adjust and stay alert, and as a parent myself who had religiously read the parenting books and attended the birthing classes and done the pregnancy exercises and bought all the unnecessary baby gear, I felt a pang of recognition: He was trying so, so hard.

Nearly a decade after her last trip to Borneo, in the fall of 2022, Johana was walking endlessly around Singapore, brimming with frustration. She'd been trying to get back to the rainforest for years to search for more evidence that the frogs represented a true case of, for lack of a better term, sex-role reversal. (She acknowledged that the phrase was problematic, but since no one had come up with a better one yet, "this is what we've got right now," she said.) In the latest of several obstacles, she was having trouble getting a travel visa to the Malaysian part of the island—where she'd wanted to scout other locations that the smooth guardian frogs might inhabit—because of strict entry rules for Colombians. "Not all

Colombians are criminals you know, some of us just want to do our job," she wrote to me. Johana thought, with some bitterness, that Americans could waltz in and out as they pleased. Sometimes, on past expeditions to other countries, her American research assistant had breezed through while she, the project leader, had been held back.

After her 2014 fieldwork trip, she'd wanted to observe the frogs' mating process, whether the females ever fought with each other, and whether the males chose mates based on certain criteria such as the females' calling frequency. "This is my baby," she said. "I want to keep working with them until I retire."

But she had no money to continue her research; after she got funding, the pandemic hit; and after travel reopened, she ran into delays getting her research permit approved. "I've been trying to get back to Borneo for so long now," she said. "I don't even remember the forest." She worried that the smooth guardian frog population had been decimated: "What if I go back and they don't call?" Unable to get her Brunei permit in time, she'd switched to plan B, to look for frogs in the Malaysian part of the island.

Now, stuck without a visa, Johana ended up going to the Philippines to help another scientist do unrelated fieldwork instead. And a last-ditch attempt to get permission to search for smooth guardian frogs in another part of the Philippines didn't come through in time. She had thought this would be an epic expedition, and she was returning to the United States having accomplished none of what she'd planned. "Empty handed. I'm flabbergasted," she wrote to me. "I feel heartbroken and defeated."

When I'd asked other biologists about the research by Johana's team, the general consensus was that smooth guardian frogs did seem unusual. Some of their behaviors had been observed in other species; for instance, male coquí frogs sat with eggs for a few weeks and mostly stopped seeking new mates during their parenting duties, but the females didn't appear to call during courtship. In Majorcan midwife toads, the females sometimes called for mates and grappled, and the males carried the eggs, but the males didn't devote their attention exclusively to one clutch.

"We know that some other species tick some of the boxes," Fischer said. "What's important about their study is that they ticked all the boxes."

But Johana would need to do more to demonstrate sex-role reversal. For example, she'd have to perform experiments to find out if males preferred female calls with certain characteristics and whether females with those calls had more offspring. Or she would have to show that females competed with one another more often than males did. "In reality, it's really, really hard to prove," Hare said. "Especially in this animal," which sounds like "a nightmare to study."

After returning from her failed trip to Singapore and the Philippines, Johana took a leave of absence before starting a new job as an assistant professor at the University of Missouri in Columbia. She wanted to spend time with her family, whom she'd seen only sporadically over the last several years. Still, she knew she would return to Borneo. Sometimes she woke up at night wondering, had she really heard the female frogs calling, or had she imagined it?

In January 2024, Johana stepped onto the rainforest trail at the Brunei field station for the first time in a decade. She'd finally made it: She had the permit and the visa. Right away, a baby smooth guardian frog jumped in front of her, a wee creature of only a few millimeters. She thought, *This must be a good omen.*

But the forest had changed. Over the last few years, a viral disease called African swine fever had swept through Borneo and decimated wild bearded pigs in parts of the island. In the past, the pigs had dug wallows in the dirt, snuffling with their snouts and creating muddy pools where smooth guardian frog tadpoles could grow. The frogs could still use puddles and ponds after rainfall, or buckets of water left out by scientists, but Johana wondered if the wallows' disappearance would make it harder for them to find good spots for their tadpoles.

And the weather was incredibly dry. Joremy anak Tony, a research assistant from Universiti Brunei Darussalam who is a member of the Iban tribe, had been surveying frogs in the area for nearly a decade, and he had never seen such a long stretch of rainless days. The leaves crunched under Johana's feet, and it was so hot that sometimes she couldn't walk barefoot on the field station's wooden walkways. Because there was no rain, the frogs weren't calling: They and their offspring needed moisture and pools of water to survive, so they weren't going to seek mates and reproduce in this weather. In the past, when Johana had played a male call on speakers in the forest, she'd find two or three females gathering around. But this time, almost none emerged; after a dozen tries, she saw only a couple of lone females, never more than one at a time.

All of this was blowing up her ambitious plans to investigate sex-role reversal. She'd wanted to dust female frogs with different colors of fluorescent powder, record their calls, and observe which female a male chose based on the color that rubbed off onto his skin. But now she couldn't even find enough frogs to conduct her experiments.

At the very least, Johana wanted to see the mating process, which she hadn't observed on her previous field expeditions. With Joremy's help, Johana occasionally managed to capture a male and female pair calling to each other. She put them in butterfly cages and waited. But nothing happened: The frogs jumped around for hours trying to escape. In other species she'd seen, male frogs were eager to mate, sometimes even trying to clasp a scientist's hand or boot. But the male smooth guardian frogs weren't like that. She just didn't understand what they wanted.

Johana decided to go back to using the glass terraria. Maybe, she thought, the frogs would like to have little "houses" of leaves to mate under. So she and another research assistant placed a few curled leaves in the back to get males in the mood. When they put a pair of frogs in the terrarium, the female called to her mate over and over, but the male never responded.

Finally, the next day, Johana saw the frogs having sex.

Around 5 a.m., she found the male frog clasping the female, a position called amplexus. Not wanting to disturb them, she left a camera running and came back five hours later to find the male sitting next to a clutch of eggs. On the video, she saw the male on top of the female with his hands on the ground, as if he was doing a push-up. Then the frogs slowly started rotating their bodies in unison. Every five minutes or so, the pair spun about a quarter turn. After a while, the female released some eggs, and the male thrust his pelvis, presumably fertilizing them. They repeated this step again and again, with only a few eggs each time. Johana had no idea why they were spinning, and why the female didn't release all her eggs at once.

But the most puzzling part was that the female kept calling during the entire mating process, which lasted about an hour and a half, and she continued even after the pair had separated. Johana couldn't understand why. If she already had a mate, why didn't she stop calling after laying her eggs? Was she already trying to attract her next mate? Did she have more eggs in her body that she wanted to release?

When Johana described her findings to me, after returning to the States, she kept repeating her questions, like turning a stone over in her palm: "I saw that, and I was like, she's calling, she's still calling. Why is she still calling?" she said. This was one of the first research goals that she'd set off to Borneo with in 2012—to find out why the female called—and now she had more questions than ever. "Why is she still calling?" Johana said again. "I don't understand why she's still calling."

During our conversation, Johana and I became engrossed in analyzing every movement in her video of the mating process: where the male placed his hands, when he started moving his pelvis, who was spinning whom. While I recognized the absurdity of poring over a frog sex tape, seeing the frogs' intimate encounter also made me realize how fully these fragile creatures lived in their own world. They were absorbed in finding mates and having young and protecting them from predators and carrying them to the place where they'd grow up, while human parents in the parallel world of my suburban neighborhood were absorbed in pairing off and having kids and holding their hands across the street and walking them to school, where we dropped them off like tadpoles in a pool. All of us engaged in the business of moving life forward, whatever that might mean on our uncertain planet. And when I watched Johana's video of a female frog calling, fingers splayed, her body lurching a bit every time she *bleep*ed, I saw that she was, in her own way, also trying to turn the wheels of life's great machinery—that with her continued, insistent calling, she was insisting on the survival of her species.

During Johana's trip, she had told me about a moment with another female frog about a month into the fieldwork. One night around 7 p.m., she and Joremy heard a female call in the rainforest from far away: *bleep*, *bleep*, *bleep*. They followed the sound until they found her sitting on a leaf propped up by twigs above the ground.

After Johana recorded videos of the frog, she and Joremy decided to sit and watch the female for a while. It had rained recently, and everything looked beautiful and shining and new; after so many days of dry weather, the forest had come alive. The crickets buzzed, *krrr-krrr-krrr*; Bornean horned frogs honked, *mehp…mehp…mehp*; Wallace's flying frogs warbled nonstop for mates from a nearby pond, *brr-rrr-rrr-rrr-bup-bup-bup*. Still, it was calm enough that they could listen to the *tap-tap* of raindrops and hear their own thoughts. Johana and Joremy whispered that with most other frogs in the world, you wouldn't be able to locate females by their calls because they were silent. People might think this frog was just a little brown blob, but it was unique. As they talked, the female frog stayed on the leaf, seemingly unperturbed by their presence, calling. And Johana stayed still on the ground, wondering.

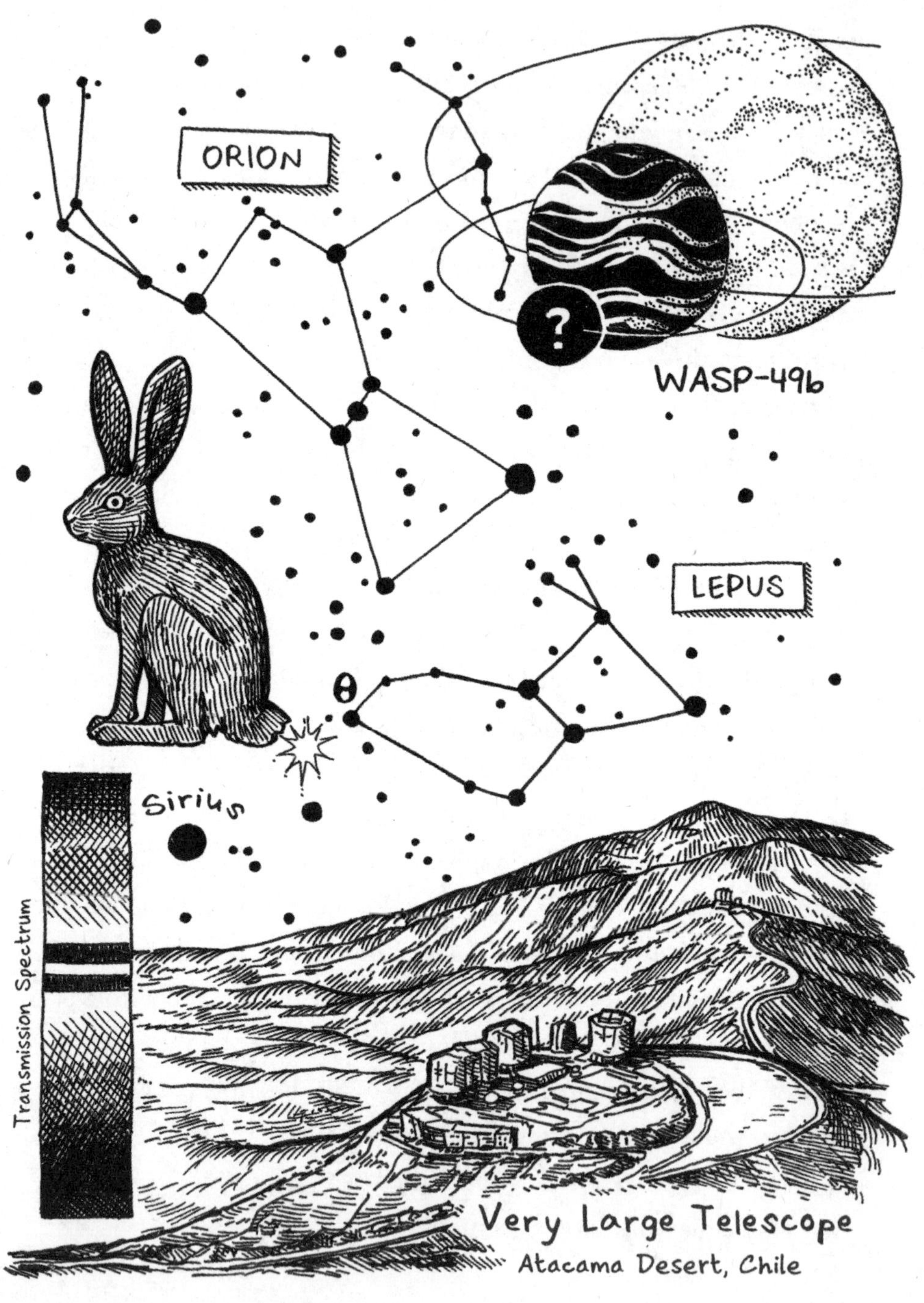
ORION
WASP-49b
?
LEPUS
θ
Sirius
Transmission Spectrum
Very Large Telescope
Atacama Desert, Chile

ASTRONOMY

Moon on Fire

"There is no signal. You should stop talking about the signal that used to be there. There never was a signal."

On a cold December night, at around 8:30 p.m., I bundled up in a puffy winter coat, knit hat, gloves, and boots and hauled a black backpack to my car. The backpack contained an amateur telescope, a gift given to my daughter years ago, which I had only a murky idea of how to operate. I drove to her elementary school a few blocks away—a site I'd picked because it might have a better view of the stars—and set up my gear in the middle of the dark, empty soccer field.

At 9:15 p.m., I called Apurva Oza, a planetary astrophysicist at the Jet Propulsion Laboratory in La Cañada Flintridge, California. He studied moons, and he told me that I might be able to see Jupiter's moons even with my kiddie telescope. Oza (he prefers to go by his last name) was now stationed at a canyon dotted with palm trees and cacti, under a clear desert sky. Here in Washington state, the clouds had started to roll in, but I could still see Jupiter, a lone pinpoint shining bright in the southwest.

"Did you see any dots?" Oza asked me on the phone. One moon, Io, was behind Jupiter, but three others—Europa, Ganymede, and Callisto—should be visible in a neat row. When Galileo had first seen Jupiter's moons in 1610, he had initially thought they were "three little stars," but "they nevertheless intrigued me because they appeared to be arranged exactly along a straight line..." I squinted through the telescope's eyepiece.

"I mean, *maybe* I see something?" I said. If I stared hard enough, I thought there might be a faint—very faint—white speck next to the bright planet.

"You should see two more specks, the same brightness as that first speck," Oza suggested.

"I don't know..." I couldn't see the second and third dots, and the first dot seemed to flicker in and out. This, I guessed, was a small taste of what astronomers felt when they tried for years to observe a celestial object; I'd been looking for only a few minutes, and now I really, really wanted to see these moons. "Astronomers spend their lives, their entire lives, searching for things that they may never find," Oza had told me earlier. "Sometimes it's sad, but I think it's also happy."

I started making excuses. "It might be that the telescope is not powerful enough," I said. "Also, it *is* still kind of cloudy here..." Then the single speck resolved a bit more clearly. "Oh no, no, no," I said, getting excited. "I *definitely* see one!" I even thought I could see a second dot. Or did I? Maybe I was imagining it into existence. Maybe there was something in my eye. Maybe... Then someone turned on a blinding light at the school, destroying my night vision. "Dammit!" I said. "Arrrgghhh!"

Oza had been trying for years to see a moon—one much farther away than Jupiter's. He was looking at exoplanets, planets outside our solar system. And he wanted to see an exomoon, a moon orbiting one of those faraway planets.

Scientists had started spitballing in the late 1990s about the possibility of detecting exomoons. David Kipping, an astronomer at Columbia University in New York City, labored for years to find subtle variations in starlight that would suggest the presence of these objects, and over the last decade, his team had identified two possible moons around exoplanets. Other researchers had turned up intriguing hints—a curious glow or the bending of starlight around a potential moon. But the scientific community hadn't accepted any of these reports as fully convincing. And numerous other cases, which had seemed promising at first, had been discarded or dismissed. Typical news headlines describing developments in the field included "Have astronomers detected exomoons at last?" and "Astronomers have spotted 6 new moons in other planetary systems! *Or have they*?" (italics mine).

"We've been sort of stymied," said Alex Teachey, an astronomer then at the Academia Sinica Institute of Astronomy and Astrophysics in Taipei, Taiwan, who worked with Kipping. "People expected more by now."

Exomoons hold fascination partly because they might harbor

life. One of NASA's goals is to find planets similar to our own elsewhere in the universe, but "surely there's these two pathways of having Earth-like worlds," Kipping said. "One of which is planets, and one of which is moons." And a moon might remain habitable longer than a small terrestrial planet would, said Michelle Hill, an exoplanet astronomer now at Stanford University. A moon has extra sources of heat that could create the conditions for liquid water, such as constant friction caused by the planet's changing gravitational pull and reflected light from the planet. "It's also getting a little bit of extra shine," Hill said.

For the public, the idea of lunar life often conjures up romantic sci-fi visions of adorably furry Ewoks on the forested moon of Endor in *Return of the Jedi*, or the lush illuminated landscape inhabited by the Na'vi on Pandora in *Avatar*. And astronomers aren't immune to these Hollywood fantasies either. "We're a bunch of nerds," Hill said. "Of course we want *Star Wars* to be real."

For Oza, though, his desire to find an exomoon had nothing to do with extraterrestrial life. The moons he sought were hot, sputtering, volcanic worlds blasting jets of lava, some on the verge of disintegrating. (He still leaned on *Star Wars* references, describing one possible moon as a place where Jedi go to die, "perilously familiar to Anakin Skywalker.") Oza's strategy was to find clouds of gas evaporating from moons, a signal that he called, appropriately, the "smoking gun."

Even though these volcanic worlds wouldn't be habitable, finding the first exomoon would make a lasting mark on the rapidly growing map that astronomers were drawing of worlds outside our solar system. The hunt for exoplanets, which took off in the 1990s, was "the dawn of an age of discovery," wrote one astronomer in a blog post on the *Scientific American* website. Scientists were seeing an immense new landscape unscroll before them. A pair of researchers who identified one of the first exoplanets had received the Nobel Prize; while exomoons might not carry quite the same cachet, they had the advantage of still being undiscovered. Thousands of exoplanets had been cataloged, but their moons remained out of sight, like distant islands off the coast obscured by fog.

And they presented a tantalizing challenge: Teachey once described exomoons as "among the most elusive objects in observational astronomy." Finding one would be akin to spotting a giant squid thousands of feet below the surface of the ocean, or a secretive, never-before-seen bird in the depths of a remote forest. When one researcher had asked Kipping, "Why not just look for hot Jupiters?"—a type of exoplanet, of which astronomers have identified hundreds—he found the suggestion utterly uninspired. "Who cares about hot Jupiters?" he thought; anyone could find those. For Oza, the difficulty of the search was partly what made it so enticing. "The status quo is, 'Don't search for this because it's like looking for a tiny grain of sand in an ocean,'" he said. "It's like a forbidden fruit."

But the danger of such a search was the powerful lure of believing that you'd finally found what you'd been seeking. Like spotting a shadow of a tentacle in the ocean or a flash of color in the treetops, you might glimpse a flickering signal in patterns of light transmitted over vast distances to a telescope on Earth and see what you wanted to see, a sort of cosmic mirage. Hunting exomoons was an exercise in warring impulses: to throw yourself wholeheartedly into the quest and fight the human tendency to declare victory too early. Over and over, Teachey had thought that he might have a discovery on his hands, and then "it falls apart," he said. "You can't really get worked up."

During my time following Oza's work, I'd read an account of scientists who had observed an odd plume above one of Jupiter's moons in the 1970s. They wondered if they'd found something momentous, a phenomenon never before seen in space, and they were trying to remain calm. But "the more you work with it," one scientist remarked, "the more you think it's real."

I had first talked to Oza in April 2021, after coming across a paper that his team wrote about their exomoon search strategy. I asked Oza if he was still planning to look for moons or if he was done with that part of his research, and he sounded fired up. "Definitely not done," he said. "I'm still searching." He'd pinned his hopes on a planet with the unfortunate name of WASP-49 b, which sounded less like a mysterious faraway world than a cross between a pesticide and WD-40. About 3,700 trillion miles away, the planet orbited a star in the constellation Lepus, the hare chased by Orion.

Oza had curly shoulder-length hair, a neatly trimmed mustache and beard, and oversized gold-rimmed glasses. He peppered his emails and texts with exclamation marks and emojis; in conversation, his stories tumbled into many directions like rivulets of water spilling through every opening in a riverbed, unable to contain themselves. When I asked how he'd gotten interested in science, hoping for a linear chronology of his life, he described an image he'd seen in second grade of Neptune, being inspired by the planet's rich azure color… and then seeing that same shade of blue in an abstract painting by the French artist Yves Klein at the Centre Pompidou in Paris, where he'd lived in his twenties during his PhD studies… how that Neptunian blue made him think of oceans, sailing far across the sea…or the time he'd seen a picture taken by the Hubble Space Telescope in eighth grade, capturing galaxies as they appeared billions of years ago… being absorbed by the image's sheer beauty, gleams of a glamorous yellowish orange…which would turn out to be the color of the element he'd start chasing in graduate school: a sodium signal from volcanic moons, which evoked images of lights in classical cities like Paris and Milan—the signal he was still seeking now, full circle to eighth grade…

Oza made frequent reference to his cultural heritage (when he'd visited the library during his early childhood in Princeton, New Jersey, "my Indian immigrant mom was like, 'You'd better get a science book,'" hence the Neptune picture) while also weaving in threads of the French world he'd been immersed in during graduate school. He sketched bhangra choreography in his research notebook; pronounced my name with a French accent ("Ro-*BEAR*-tah"); drank chai from a pint glass ("Is that allowed?" he'd asked an Indian collaborator during a videoconference); and had posted a 1969 article from *Le Monde* on his website, the headline proclaiming "Deux hommes ont foulé le sol de la lune" (Two men set foot on the moon). During his graduate studies, he'd created a Dance Your PhD video with other students—an exuberant mash-up of Indian folk and hip-hop dancing, meant to represent Jupiter's moons and oxygen molecules, filmed along the

Seine River. In one part, they'd thrown blue and red powder, used in the ancient Indian art form rangoli, to enact lunar volcanic eruptions.

Oza's interest in exomoons was first piqued in 2012, when he was a master's student in astrophysics at the University of Virginia in Charlottesville. His adviser had told him to talk to Robert Johnson, a scientist then in his seventies; Robert and a colleague had come up with an exomoon search strategy inspired by what researchers had learned over the last several decades about Jupiter's moon Io. He was a molecular physicist, not an astronomer, and had initially assumed that astronomy was just about dead objects. "Who cares?" Robert said. "Then it turns out that Io is not a dead object."

On March 2, 1979, astronomers had published a prediction that Io was covered by active volcanoes. Their reasoning was that Io and two other moons, Europa and Ganymede, always passed each other at the same points in their orbits, exerting extra gravitational tugs every time. This constant tugging had pushed Io into an oval-shaped, rather than circular, orbit around Jupiter. So sometimes Io was closer to the planet, sometimes farther away.

Jupiter's gravitational forces, which stretched the moon, were thus sometimes stronger or weaker depending on Io's position. The constant changes in stretching heated up the moon, melting its innards into magma. As a result, "one might speculate that widespread and recurrent surface volcanism would occur," the authors wrote. And the spacecraft *Voyager 1*, which would arrive at Jupiter three days after the team published their report, might be able to see if their predictions were correct.

One member of the *Voyager 1* team was Linda Morabito, an engineer at the Jet Propulsion Laboratory (JPL). She worked in an office that constantly blasted cold air to keep a computer cool "in a cacophony of horrendous noise," she wrote in an account of her research on the mission. On March 9, while analyzing a new picture of Io that she had initially dismissed as useless, Morabito saw a bright semicircle peeking from one side of the moon. "Jesus, what's that?" her colleague said.

The colleague had heard about the recent prediction that Io was volcanically active, and other JPL researchers wondered if the blob was a cloud of gas. After further investigation, the team realized that the semicircle was located roughly above a mysterious heart-shaped mark on Io. Mission scientists decided to turn on a spacecraft instrument to look for more evidence and to examine other images more closely. On March 12, one of the astronomers called Morabito and said, "You'd better get up here! They've found volcanoes all over the place!" She could hear people screaming. It was confirmed: They had found the first evidence of an active volcano in outer space. The semicircle was in fact "an enormous umbrella-shaped plume," the *Voyager* team later wrote, erupting about 165 miles into the air.

Robert Johnson had followed these developments with interest, and he knew that one element released in Io's eruptions was likely sodium. In 1972, an astronomer had observed "an anomalous brightness" due to sodium in the light coming from the moon. And after the *Voyager 1* mission, scientists cataloged even more dramatic sodium signals at Jupiter and Io.

Then, in 2002, astronomers reported that they'd seen sodium at a planet outside our solar system, circling a star in the constellation Pegasus. "I thought, oh my goodness, maybe this is something that's happening there that's related," Robert recalled.

Those researchers assumed the sodium was from the exoplanet's atmosphere. But what if it was coming from an exomoon orbiting the planet instead—a moon spewing volcanic gases, like Io? In a 2006 paper, Robert and a colleague reasoned that these types of signals could be a way to find moons too small to observe by other means. Sodium happens to be easy for astronomers to detect, compared to other elements. So if the moon blasted a sodium cloud into space, "you could see that even if you couldn't see the tiny moon," Robert said.

When Oza showed up at his office six years later and Robert explained the idea, it made complete sense. "It's so real that it's a worthy chase," Oza told me. And no other scientists had put forth a persuasive exomoon detection yet. Oza "just knew that if he was the first one to prove there was a moon there, he would be well-recognized," Robert said.

When I asked Oza if he was driven by that desire to be first, he demurred. "It's more about the excitement of once you find them," he said. "To know that they're there, for yourself and to share that with others." Still, he acknowledged that his upbringing as a child of immigrants had trained him to push limits, and he probably had a built-in tendency to try to find something new; his first name, Apurva, means "never before."

This feeling of being the first was eloquently described in Morabito's account of seeing the unexplained semicircle behind Io. Left alone with the mysterious picture, "I knew I had something here," she wrote. "I sensed that I was seeing something that no human being had ever seen before." Her heart was pounding. This moment was "the only time that has ever belonged to me and this discovery," she wrote. "It was the stuff of dreams."

Around the same time that Oza met Robert, Kipping was launching the first intensive search for exomoons. He'd done some theoretical work exploring how moons could be detected and finally decided to "start looking for these things." When I'd spoken to Kipping during that early phase, he sounded optimistic and excited; *Popular Science* picked him as one of its Brilliant 10 and ran a glowing illustration of Kipping grinning, dark hair lit with a bluish Superman-like tinge in front of a backdrop of stars.

But if Oza was the bright-eyed recruit, Kipping would eventually become the world-weary veteran. On his lab's YouTube channel, Kipping described, with unusual frankness, an emotional rollercoaster of highs and lows. "Each and every search has come with a fresh lining of hope, a wistful skip in our stride, as we dare to imagine that this time—despite all of the failed searches of the past—*this* time might be different," he told viewers in one video. "But time after time, I've been left with another disappointment." Teachey, one of Kipping's former students, described the process as akin to running a marathon with an egg on a spoon. "You get one meter from the finish line and the egg drops," he said. "And you've got to go all the way back and pick up another egg."

Kipping had started thinking about exomoons around 2008, when he was a PhD student in London. Other astronomers had speculated that exomoons could be detected using a strategy called the transit method. As an exoplanet passed, or "transited," in front of its star, it blocked some starlight. So if scientists measured the amount of starlight reaching the telescope, they would see a dip in light intensity. If a moon was present too, plonked on the side of the planet, it also should block a bit of starlight and cause an additional dip.

Scientists could look for other clues too: For instance, the moon might change the timing of the planet's transit. The moon's gravity would tug on the planet, making it wobble. Instead of orbiting the star like clockwork, the planet would transit a bit earlier or later than expected.

Kipping felt sure that exomoons were out there. "It's kind of ridiculous that there'd be no other moons," he said. "They clearly do exist; it's just finding where they are."

It was the "finding where they are" bit that was tricky. When I spoke to Sébastien Charnoz, an astrophysicist at Université Paris Cité in France, he ticked off a long list of reasons why finding an exomoon was so hard. First, moons were small compared to their planets. So the dip in starlight caused by a moon would be minuscule—so small that it would be difficult to distinguish a real signal from an instrumental artifact or random activity on the star's surface.

To make a solid case, you needed data from multiple transits. But the moon might not cause a light dip every time the planet transited because sometimes it was behind the planet. "You see the problem?" Charnoz said. "When you do not see anything, it does not mean there is no moon. It may mean that the moon was hidden."

To increase your chances of seeing multiple moon transits, you could focus on planets that circled their stars very tightly and thus transited more often. But planets that were very close to their stars were also less likely to have moons at all, partly because a moon could more easily be pulled away by the star's gravity. And if you studied planets that orbited farther from their stars, and thus were more likely to have moons, you'd have to wait much longer for each transit—sometimes years.

The seeming impossibility of the search reminded me of French astronomer Le Gentil's doomed quest to observe Venus crossing the sun in 1761 and 1769. Le Gentil embarked on voyages to India and the Philippines to observe the transits of Venus, which occur in pairs roughly once a century. He was foiled by war, uncooperative winds, and a hostile government official on his journey and ended up unable to collect any usable data on both transits—the first because his destination in India was captured by the British, prompting the ship's crew to turn back, and the second because "a fatal cloud" obscured his view. "I was more than two weeks in a singular dejection," he later wrote of his failure, "and almost did not have the courage to take up my pen to continue my journal; and several times it fell from my hands, when the moment came to report to France the fate of my operations…"

Kipping didn't fare much better than Le Gentil in his early search. In 2009, NASA's Kepler telescope launched into space and started capturing transit data on thousands of potential exoplanets. Kipping, who had moved to the Harvard-Smithsonian Center for Astrophysics in Cambridge, Massachusetts, formed a team to look for moons around some of those planets. While he was on his honeymoon in Mauritius in 2011, he got an email from a team member about an exoplanet with "a big beautiful exomoon-like signal," Kipping said; further digging revealed a "huge" change in transit timing. But their excitement was short-lived. One of his colleagues figured out that the change in transit timing was due to gravitational tugs from another previously undetected planet.

And the dip in light was caused by a starspot, a patch on the star's surface that had temporarily cooled.

Kipping kept looking. A couple years later, another intriguing maybe-moon showed up in observations of an exoplanet with "an amazing transit signal," he said wistfully, accompanied by an extra dip during several orbits. But two of the moonlike signals were caused by an instrumental oddity that led to data corruption. When the team realized what had happened, Kipping was already writing the report of their discovery. "I was like, oh fuck," he said. "We can't publish this."

His hopes were dashed again when his team investigated a possible moon reported by scientists in Europe in 2014. The signal turned out to be caused by misbehaving electronics: A blast of high-energy radiation, perhaps a cosmic ray, had reduced the sensitivity of pixels at the exact spot in the light data where a moon could be orbiting. "Ridiculous coincidences," he said. Another exomoon killed.

When Kipping told me these stories over the phone, he had a memory of sitting on a bench by a pond after finding one of these near-moons. He'd had so many promising cases at that point that he couldn't even remember which planet he'd been studying at the time. He felt giddy with the thought that this might be the first exomoon ever detected; taking a break on the bench, he tried to calm down and be more objective. He knew he had to guard against what he wanted to be true. But he was thinking, "This could be the one."

In 2013, Oza started searching for sodium clouds around Jupiter-like planets at the Apache Point Observatory in New Mexico. On his first attempt, the view was blocked by clouds. When he tried again the following year, he had to wait out a sandstorm and finally got a clear window later in the night. But when he examined the data, he couldn't find a sodium signal. "I tried my heart out," he said.

Like other exoplanet astronomers, Oza was using a technique called transit spectroscopy. When a planet passed in front of its star, the starlight filtered through the planet's atmosphere before reaching the telescope. Atoms and molecules in the atmosphere absorbed starlight at different wavelengths, essentially "catching" the light and preventing it from reaching the telescope.

Scientists could then measure the intensity of light at each wavelength, using an instrument called a spectrograph. If the light level at a certain wavelength dropped, that could mean a particular element or molecule was present in the planet's atmosphere. Sodium absorbed light in the yellowish-orange part of the spectrum, the color that had entranced Oza as a kid.

After his attempts at the New Mexico telescope failed, Oza didn't have any more funding to continue the research. He gave up temporarily and began a PhD program in Paris, studying Jupiter's icy moons Ganymede and Europa. I asked Oza if he stopped thinking about exomoons during that time. "Believe me," he said, "I couldn't stop thinking about them." He worried that other scientists would detect the signals he wanted to find before he had a chance to look again, and he talked only to a few close colleagues about his exomoon search strategy. "I held onto it like the biggest, darkest secret ever," he said.

In 2017, a new report caught Oza's attention. A scientist

named Aurélien Wyttenbach, then at the University of Geneva in Switzerland, and his colleagues had observed sodium at the planet WASP-49 b, a Saturn-like gas giant made of hydrogen and helium. Sodium mostly absorbed light at two wavelengths, which caused two dips in the spectrum, or a "doublet." The team had detected the separated doublet using a high-resolution spectrograph, and the dips were spectacular. Compared to what scientists had seen at other exoplanets, the sodium signal at WASP-49 b was huge.

Wyttenbach's team had concluded that the sodium came from the planet. But the signal's unusual strength bothered Oza. To explain the results, the sodium had to be present more than twenty-five thousand miles above the planet's surface. The altitude was so high that Oza decided, "This has to be impossible." Based on his team's computer simulations, it seemed more likely that the sodium came from a moon instead.

At one point in our conversations, Oza made an oblique reference to a "long and dark road." He told me that right before his PhD program ended, on his twenty-seventh birthday, he found out that his best friend, Tony, had died by suicide. During his master's program, Oza had told his friend about the exomoon idea, and Tony had asked smart questions, thrown axes at the theory. I asked, stupidly, how Tony's death had affected him.

"I would say, really terribly," Oza said. He didn't know how to grieve. Looking back, he said, he was probably depressed for parts of the next two years. Oza moved to Switzerland for a postdoctoral position at the University of Bern and kept pursuing the exomoon idea. Some people advised him to work less intensely. But when his mind wasn't occupied, he felt like he was in a bad place. To be honest, he said, "exomoons kept me alive."

On a December night in 2020, an astronomer named Jonathan Smoker was stationed at an observatory control room atop a desert mountain in Chile. A large computer screen tracked weather conditions such as wind, clouds, and humidity; outside, the rolling reddish-brown landscape was stark and devoid of life, which had made it an appropriate site for a James Bond villain's hideout in the film *Quantum of Solace*. Smoker worked at the Very Large Telescope, which actually included four very large telescopes with twenty-seven-foot-wide mirrors. One of them was pointed at WASP-49 b.

About 4,200 miles away, Oza followed the observations from an Airbnb rental in North Carolina, where he was temporarily staying near his family. (In the modern era of astronomy, scientists often don't travel to an observatory; staff members on-site operate the telescope, perform the observations, and relay data.) A few years had passed since Wyttenbach's report, and Oza had decided to observe WASP-49 b for three nights to look for stronger evidence of a moon. He'd use a bigger telescope and a higher resolution spectrograph than Wyttenbach had, which might allow him to pinpoint the source of the sodium.

On the first night of observations, everything went smoothly. But on the second night, about a month later, the team ran into a technical problem with the spectrograph. They switched to another telescope and the data got corrupted. By that time, WASP-49 b's transit across its star was over.

Oza prayed that the third night would work. This time, the telescope couldn't be pointed at WASP-49 b because of strong winds, which might blow dust onto the mirror. Oza, now stationed in his parents' basement in Raleigh, spiraled into melancholy. In astronomy, new discoveries hinged on being awarded time at powerful telescopes, but these applications were highly competitive; there was no guarantee he'd be given another chance. Eventually the winds died down, but Oza managed to catch only part of WASP-49 b's transit.

Later that year, I joined Oza on a video call as he met with an exoplanet astronomer named Julia Seidel to talk about the data. "Thank you, ESPRESSO, for fucking up our observations, right?" Julia said good-humoredly. (ESPRESSO was the name of the spectrograph, which apparently caused consternation for astronomers who tried to search for guidelines online and found only coffee machine manuals.) Julia, one of Oza's collaborators, had just moved to Chile to work at the European Southern Observatory in Santiago and was sitting in a bland, beige-toned hotel room. She suggested to Oza that they could apply for a night of observations using the combined light of all four telescopes, which would have "ridiculous" precision, she said. "But it's really hard to get. So you need to have a waterproof science base."

Julia seemed to view science as an elaborate strategy game, acting as both an unflappable sidekick to Oza's brooding Batman and the hard-ass who mapped out their next moves with clockwork precision ("because I'm the German"). She once described how she'd swapped data sets with another team for mutual benefit, "like a prisoner exchange," and told Oza a cautionary tale about an astronomer who hadn't published his study quickly enough; another team had grabbed the data when it became public and reported the results first. "He got scooped hard," she said. "He was really sad about it." When she discussed the internal politics of astronomy, they sounded like delicate UN diplomacy situations. "I don't think that this is happening, neither from the Canary Islands nor from Leiden," she said at one point, referring to a breach of scientific etiquette that went over my head. "They're way too scared of the retaliation from Geneva."

During the ill-fated second night of his WASP-49 b observations, Oza had written to Julia bemoaning his fate. "You were very upset," she later said.

"I was emotional, yeah." This was after he'd had to abandon one telescope and try another telescope called UT3 instead.

"It's the worst of them," she said. "The moment you wrote me that you're trying to switch to UT3, I was like, this is a goner."

"Julia has incredible intuition," Oza told me.

After he'd performed the observations, Julia had analyzed the data from the first night. Oza was optimistic that they would confirm Wyttenbach's sodium observations. "That's what I kind of convinced myself, that I'll see a huge signal, bigger than ever," he said.

But after Julia had processed the data, there was no sign of sodium. She was surprised; the spectrograph was better than the one in Wyttenbach's study, so "you should see even more," she said. "There was nothing to be seen."

Over the last few years, Kipping had been embroiled in his own exomoon drama. Now at Columbia University, he and Teachey had tried a slightly different method: They'd "stacked" the transit data from about 280 exoplanets together in hopes of making it easier to see a signal. When Teachey finished the analysis, Kipping came into his office and looked at the results. "There was just nothing there," Kipping said. "I was just gutted." He felt desperate and asked Teachey to look quickly at each exoplanet's data individually.

They ended up finding a dip at a roughly Jupiter-sized planet called Kepler-1625b. And when Teachey and Kipping observed it again using the Hubble Space Telescope, they saw the two signals they would expect from a moon: a blip after the planet's dip and a change in the timing of the planet's transit, which was seventy-eight minutes early.

The moon, if it existed, was "extraordinary" and "bizarre," Kipping said in videos describing the discovery. Based on the size of the signal, it was enormous: about as big as Neptune, prompting a colleague to call it a "Neptmoon." Kipping and Teachey presented the discovery cautiously, writing that the moon "cannot be considered confirmed until it has survived the long scrutiny of many years..."

The scrutiny arrived promptly. Two teams reanalyzed the same data, and neither was convinced. One team thought that another planet could account for the early transit arrival and argued that "the exomoon hypothesis heavily relies on a chain of delicate assumptions, all of which need to be further investigated." The second team, after using different methods to process the data, couldn't reproduce the dip and concluded that there was "no evidence for lunar transit." Kipping himself felt "fifty-fifty about it," he said. "It's still not enough data to really know."

By this time, other teams had put forth potential exomoons

too. There were gaps in a disk of gas around an exoplanet, which suggested moons might be present there. There was a candidate found through a technique called microlensing: Astronomers saw a peculiar pattern of brightening in the sky, perhaps caused by light from a distant star bending around a planet and its moon. There was a blob of light that might be the glow of a young luminous "companion" circling a planet or a brown dwarf (an object somewhere between a planet and a star). But all of these candidates remained in a gray zone. The upshot, as Kipping put it in a video, was that "the field of exomoons finds itself with dozens of no's and just one maybe."

When I asked Kipping what he thought of Oza's sodium search strategy, he was supportive, but with caveats. "I think it's a cool idea," he said. (He'd talked to Oza recently; they'd met at Tom's Restaurant, the *Seinfeld* diner in New York City, which Oza had suggested because he thought it would be "iconic.") Still, he said that he liked the transit method because his team could predict when a moon signal should occur again. "There's a definitive repeatability to them that would allow us to have just a solid, slam-dunk detection," he said. "I don't see, personally, a pathway to that with sodium lines."

Oza was disappointed that Julia hadn't seen any sodium in the observations from Chile, but he was undeterred. He'd been working with other scientists who had observed WASP-49 b from another telescope, perched on the summit of the mountain Maunakea in Hawaii. Athira Unni, then a PhD student at the Indian Institute of Astrophysics in Bengaluru, had started processing the data, and this time, a sodium signal did emerge.

But Julia was skeptical. The signal was "incredibly large," she said. "Larger than I'd ever seen." Like Kipping, the team faced a long list of potential minefields. First, they had to make sure any changes in light weren't due to starspots, which caused variations in the star's luminosity. Second, they needed to show that the sodium signal wasn't coming from Earth's atmosphere. For instance, if they were "super-duper unlucky," Julia said, the light emitted from Earth's sodium could make the spectrum appear to have a sodium dip during the planet's transit.

Finally, they needed to address something called the Rossiter-McLaughlin effect, which apparently was not a political pundit show but yet another way that astronomers could be tricked. In what Julia called the biggest drama in recent years, a graduate student in Spain—whom she admiringly dubbed "the sodium killer"—and her colleagues had shown that the very first report of sodium at an exoplanet, published in 2002, likely wasn't a true signal at all. The reasoning was convoluted and had to do partly with the rotation of the planet's star causing the wavelength of light reaching the telescope to be squeezed or stretched, creating bumps in the spectrum. It was a "fantastic paper destroying one of the most famous papers in all of exoplanet science," Julia said. "People were pretty stunned."

For the WASP-49 b observations, she thought they were probably safe on this count. In their earlier study, Wyttenbach's team had concluded that this phenomenon wasn't a concern because the planet's star rotated very slowly. "It's not going to bite us in the ass," Julia said. "This can't be only the RM effect."

But in December that year, the data from the Hawaii telescope fell apart for other reasons. When Athira performed additional steps to account more rigorously for the effects of Earth's atmosphere, the sodium signal from WASP-49 b completely disappeared. "All our alarms went off," said Jens Hoeijmakers, an exoplanet astronomer at Lund University in Sweden, who was collaborating on the research. He was emphatic. "That means there is no signal," he said. "You should stop talking about the signal that used to be there. There never was a signal."

Meanwhile, Kipping had been burned so many times that he was rejecting exomoon candidates for the most minor reasons, the slightest hint that something was out of place. "Are you telling me this is the real one after twelve years? Well, surely not," he said. "You get so used to not seeing anything that it becomes almost a pattern, and you find it hard to accept evidence."

This time, his team was examining "cool giants," large Jupiter-like planets that orbited farther from their stars; these planets, they thought, were more likely to have big, detectable moons. They'd combed through transit data on seventy planets, ran an exhaustive battery of tests, and ruled out one planet after another. In the end, they'd been left with one called Kepler-1708 b: The transit data had a dip in starlight that they couldn't explain with any other reason. "I just didn't have any more tests I could think of," Kipping said. "I don't know what else to do."

The moon still wasn't a sure thing; his team estimated the probability of a false positive was about 1 percent. That percentage was "really frustrating," Kipping said in a video explaining the results, because it hovered on the boundary of what would be considered a real discovery. And again, this potential moon was gigantic: more than twice as big as Earth or about the size of a mini Neptune. "I naturally have some hesitancy about its reality," Kipping said.

I asked Kipping if he was happy about the finding. "In terms of the objects themselves—I know it sounds kind of strange, but I'm almost divorced from them," he said. "I have no feelings about them." After running so many tests, "I'm almost emotionally exhausted."

"Isn't that kind of sad?" I said.

"I don't know if it's sad," he said. "I like that, because I don't want to be emotionally attached to any object I discover." He told me about an astronomer who had nicknamed a newly detected exoplanet Zarmina's World, after his wife. Later research suggested the planet wasn't real, but for at least a couple of years, the astronomer continued to defend his team's claim. "How can he turn around to his wife and say, 'You know that planet I named after you? It's not real,'" Kipping said. "You can't. You've lost objectivity."

At the time that Kipping's report came out, I didn't know that about two years later, a team in Germany would claim to refute both of his candidate moons. The scientists reanalyzed the data and concluded that the evidence didn't statistically favor a moon explanation or that the signal could be explained by unrelated factors. Kipping's team then wrote detailed arguments to refute the refutation and dismissed the German team's reanalysis as "fundamentally flawed."

When I saw the headlines, my first reaction was, *You've got to be kidding me.* One news story ran the dramatic all-caps title "NO EXOMOONS" and, to drive the point home, included an illustration of a moon stamped out, *Ghostbusters*-style, by a red slashed circle. And so it went, around and around.

Oza was thrilled. His team had found something, he said, that was "completely bonkers."

After the observations from the Chile and Hawaii telescopes yielded no clear signals, Oza had decided to return to the data on the enormous sodium signal reported by Wyttenbach in 2017. "Take it apart, start over, start from scratch, look at it upside-down, look at it sideways," he said. Wyttenbach's team had averaged the signal over three nights of observation; Oza asked Julia if she could analyze each night separately.

On the first night, a small sodium signal appeared for part of the transit. On the second night, the signal was huge—"this extraordinarily deep, like gigantinormous cloud of sodium for about forty minutes," Julia said. Then on the third night, nothing.

This pattern, Oza thought, was compelling evidence for a moon. If the sodium was coming from the planet, it should have been present for all three nights. But with a moon, you'd expect the signal to flicker on and off. When it was in front of the planet, you'd see sodium, and when it slipped behind the planet, the sodium would disappear too. "It can just hide back there for a while," he said.

Julia was less enthused. "I'm right now in between 'We have no idea what's happening' and 'This might actually be a moon,'" she said. Their data from the observatory in Chile "shows us as good as nothing," and Wyttenbach's data was "a mixture of nothing and something."

And Jens was starting to wonder if sodium was truly present in any of these observations at all. Maybe Wyttenbach's signal wasn't real either. If that were the case, he said cheerily, "that in itself, I think, scientifically would be an extremely interesting turn of events." The Wyttenbach report was held up as a landmark study. If it was wrong, scientists would want to know.

In a video call with Julia about a month later, Oza showed her an animation of the exomoon's orbit based on computer simulations. The sodium cloud would disappear as it circled behind the planet. "I'm convinced of that," he said.

"Maybe not convinced, but at least it's a plausible explanation," Julia said. Then she countered with another argument—the opposite of what he'd proposed. When the moon was behind the planet, it was closer to the star. The star's intense heat would roast the moon, making sodium plumes evaporate from its surface. When the moon orbited around the side of the planet, that sodium signal would become clearly visible. But when it moved in front of the planet, the planet would block the star's heat. The moon's temperature would plunge, turning it into "a frozen rock for a little while," and sodium would stop evaporating. In other words, the sodium signal would disappear when the moon was in *front* of the planet, not behind it.

She held up an eyeglasses case and a cup to represent the moon and planet during the three nights of Wyttenbach's observations, moving the case around the cup. "We see something in the first, we see a shit ton in the second—very localized—and in the third, we see nothing."

"I think you got it," Oza said. "I think that is the ticket." He leaned back in his chair and put his hands behind his head, apparently thinking through the scenario. "Wow," he said, rubbing his mustache. "That is such a nifty explanation."

The moon, he envisioned, was an orangish-yellow molten ball, enshrouded in steam, reaching more than 2,000°F when it swung close to the star. Sodium blasted into space in volcanic plumes or vaporized from the surface. And the moon might not last long; the planet's gravitational forces were tearing it apart. The moon was "just being obliterated," Oza said.

Though Oza seemed confident, his collaborators expressed varying degrees of caution. "I actually think it's a moon," Julia told me. "I actually think he's right. But the problem is, it's such an outrageous claim." When I called Charnoz, who was working with Oza, he didn't sound fully convinced. "If you tell me, 'Would you die to say it's an exomoon?'" he mused. "I would say no." For Charnoz, the study was an "idea paper" presenting an exomoon search strategy. He said, hedging, "It may be, by this kind of technique, that maybe we can detect an exomoon."

A month later, the team's report of their discovery was summarily rejected. Oza had submitted it to a prestigious scientific journal, and two anonymous experts had concluded that the evidence wasn't solid enough to support the team's claims. Julia's reaction was breezy. "Honestly, it's just part of the game," she said. "If you never get a bad referee comment, you're not in science." She dismissed one critique by the first reviewer, who'd said they didn't see a strong sodium signal, just "systematic wiggles"—in other words, noise in the data, nothing of substance. "That's bullshit," she said.

Jens wasn't even fazed by the idea that the second night's signal was a meaningless wiggle. If it turned out that the signal was driven by some other unknown cause, they would still have uncovered an important result, he said. It would mean that all data of this type was suspect, these methods weren't good enough to detect signals reliably, and astronomers should be even more careful about using this technique. "I'm not so worried or sad about this," he said. It would be a different story if their analysis didn't even reveal any insight about their methodology. "Then it would really be a useless result."

In December 2022, Julia and Oza were coming down to the wire to propose more observations of WASP-49 b. They needed to make it clear that their research was high-risk, high-gain, she said on a video call, and to explain "why it has to be right the fuck now."

"The best answer I have so far is that it might disappear if it's transient," Oza said. In other words, the moon might completely disintegrate if they waited too long. Also, he said, "it's about damn time to find an exomoon."

After their paper had been rejected, the natural next step was to observe WASP-49 b on at least one more night when their orbit calculations suggested the sodium signal would be strong. The planet's star would be visible for only a few more months in the Southern Hemisphere, and it was too late to apply for observing time at the Very Large Telescope in Chile during that window. So Oza had turned to a "Hail Mary" option called Director's Discretionary Time, a special allotment given to urgent observations. After about an hour of revising, Julia was satisfied. She told Oza to give a couple sections another scrub and emphasize the big potential payoff: "*With one more transit,*" she ad-libbed, "*we can find the first exomoon and it's super-amazing and it will be a* Nature *paper and we'll get a freaking Nobel Prize for it, blah blah blah.*" She signed off with a finger waggle.

About a week later, the organization that oversaw telescope operations rejected their request. Oza forwarded me the email with this comment:

sad

reassessing with Julia on what to do…

On the phone, he beat himself up over what he could have done differently: "Could've, should've, would've." Then he said he shouldn't get "into the loop in my head" about what he could have done. According to the team's estimates, the moon was close to evaporating; it might already be gone. He wondered if it had vanished soon after the second night of Wyttenbach's observations, which had taken place on New Year's Eve in 2015. "Imagine if the December 31 observation was the last one," Oza said. "That is the last time we saw it."

Meanwhile, other teams were gearing up to find exomoons in different ways. With NASA's Nancy Grace Roman Space Telescope scheduled to launch as early as fall 2026 (as of this writing), astronomers might be able to gather stronger evidence using microlensing—observing light from distant stars bending around moons. Some scientists hoped to analyze data from the upcoming Extremely Large Telescope, currently under construction in Chile, for starlight reflected from the surface of icy moons.

And another team wanted to examine the light from planets, not from stars or moons. Mary Anne Limbach, an astrophysicist at the University of Michigan in Ann Arbor, and her colleagues planned to look for dips in young, bright planets' light when moons passed in front of them. Normally the star's glare would make it impossible to identify these signals. So they were going to focus on "rogue" planets—free-floating planets that didn't orbit a star at all, but hung out on their own in space. Since these recently formed planets were very hot, nearby moons would be warmed up and might temporarily be habitable, Limbach noted. While they likely wouldn't have time to form life before they cooled down, studying the moons would tell scientists more about the chemistry of early Earth-like environments.

Eventually, the exomoon search might offer clues to whether life was present elsewhere in the universe. Some astronomers have argued that our own moon helped drive the formation of complex organisms on Earth. By stabilizing the planet's tilt, the moon kept the climate more consistent. So if moons similar to ours are present around Earth-like exoplanets, perhaps they could boost the chances of organisms evolving on those worlds as well. Teachey doubted that a moon was required; still, he said, "if we found a really nice Earth analogue, with an attendant moon something like ours, I think we'd all be excited to take a very close look at that planet and see if life might be brewing over there too."

And such discoveries could play a small part in untangling a big question: How common is life elsewhere in the universe? Either answer is important, Teachey noted. "If it's very common, then there's folks we can make contact with out there one day," he said. If it's rare, "we're really carrying the torch not just for the Earth but for a sizable fraction of the whole galaxy. And that would be a profound responsibility."

In January 2023, I met Oza at a hipster coffee shop near downtown Seattle, up the hill from a convention center where he was attending an astronomy meeting. At a table near the back, I told Oza the story that Kipping had told me, about the astronomer who nicknamed a planet Zarmina's World after his wife. I asked if he would hesitate to name his exomoon for fear of getting too emotionally attached. (Kipping's team referred to their potential exomoons by taking the planet's name and adding the Roman numeral "i," rather than coming up with memorable names.) Oza said, about attachment, "I love the emotion. I think it's part of being human." He didn't see a problem with naming his exomoon. He thought Ghanchakkar would be a good name, a word signifying chaos, entropy, and a spiral or circle shape, reminiscent of a moon's orbit. When writing his paper about WASP-49 b, he'd named the file "Ghanchakkar Babu" after a Bollywood song; he later remembered that his friend Tony had used that name as his Vine handle, and it felt like a nice tribute.

"You wouldn't worry about the emotional attachment part of it—that it would cloud your objectivity, to name them?" I asked.

He laughed. "Mmm, no." He told me about a meditation he'd done in Sikkim, India, which had made him really understand the concept of attachment. He'd been silent for eleven days, not even making eye contact with others; he ate only twice a day and took frigid baths. He was taught to try to keep everything in balance and to strive for equanimity: not to be overly attached to, or detached from, anything.

The idea reminded me of something else Kipping had said. "You don't want to claim things which are wrong, but you also don't want to miss things staring you in the face," he told me. "It's that balance that's so hard." In the account that Linda Morabito wrote about her discovery of Io's volcanic plume, she mentioned that at least one other scientist, a graduate student, had found similar evidence of an eruption in an image before she did. But a colleague had dismissed his observations and the student had dropped the matter. He'd seen the shadow of the tentacle in the ocean, the flash of color in the treetops, but the signs had been ignored.

A few days before that winter night when I'd tried to see Jupiter's moons at my kid's school, Oza had just gotten the news that their observation request was rejected. Our plan was to look at the part of the sky where WASP-49 b was located, even though its star was far too dim for me to see with my amateur telescope. I wanted to situate it in space, to close the unfathomable gap between me and this distant world a little bit.

After my brief tangent looking for Jupiter's moons, I turned to Orion, which loomed above the evergreens surrounding the soccer field. WASP-49 b's star was close to Lepus the hare's tail, slightly below one of Orion's feet. But the haze had thickened, and Orion's foot was now under a cloud.

"Well, that's where WASP-49 is," Oza said, on the phone from California, "right there." If we traveled there by spaceship, he said, we'd see a yellowish star out the window, circled by a reddish-orange planet. Orbiting the planet would be a molten moon, lava falling "in fountains that are maybe something like…unimaginable, really," Oza said.

Then he said suddenly, "Oh wow, I just saw a shooting star." He paused to admire it from his perch at the foothills of the canyon. "I'll let you wish on that one, since you're here out in the cold."

"No, you should wish on your moon!" I yelped, completely abandoning journalistic objectivity. "You should wish for a chance to see the moon again! Don't throw away your wish!"

"Okay, okay," he said. "I'll keep that one."

Motion-activated
camera trap

ECOLOGY

The Heroes We Need

**"You've become part of a bigger universe.
You just don't know it yet."**

Blake Garden, Kensington, California

In a black-and-white video, a coyote halts on a garden path, turns, and looks toward the camera with headlight-bright eyes. Another coyote trots past at a quick clip. The first animal follows with a slight limp, and its companion glances back, waiting. They both disappear into the darkness.

**Hastings Natural History Reservation,
Monterey County, California**

A skunk approaches a chair on a patch of straggly grass with its poufy tail straight up like an exclamation point. The animal stops short and darts backward, keeping its eyes on the chair. Then it pauses, blinks, and runs off.

Residential Yard, Oakland, California

Three raccoons amble into the frame, all bulbous bodies and fat, stripy tails. Two of them discover a plastic cylinder chained to the ground and start nosing about. One grabs the object with its paws and hugs it close; the third raccoon, which had been investigating a large open bin, starts paying attention. After more sniffing and pawing, one animal grabs the cylinder while another pulls a rope attached to the cap on the other end. The cap pops off.

Oyster Bay Regional Shoreline, San Leandro, California

A young human in a navy T-shirt, mustard pants, and a cap with "Hermanos" printed on the front kneels by a pair of slender trees. He secures a camera to the trunks with a strap, wipes the traces of his scent off the metal box, and trims the grass in front. Satisfied with his work, he leaves a sign stapled to the tree that reads, "PLEASE DO NOT DISTURB THIS EQUIPMENT."

Burke Museum, Seattle, Washington

Two humans, this time in a lab. A coyote carcass lies curled before them on a table. They weigh the animal on a scale, measure from the nose to the tip of the tail, cut open the body. They save pieces of the lung, the liver, the kidney, the fur, the entire heart. They pull out the intestines, a long, twisted, nubbly pinkish-gray mass, to be searched for parasitic worms.

Whidbey Island, Washington

Inside a coyote's small intestine, tapeworm larvae grow into adult worms. Round suckers latch onto the sides of the gut; as the worms mature, segments in their bodies become filled with eggs.

The coyote eats and poops, and the segments at the ends of the tapeworms detach and dissolve, releasing eggs into the feces.

Bernal Heights Park, San Francisco, California
A group of humans searches for coyote feces…

Mulford Hall, University of California, Berkeley
On a spring day in 2022, I showed up at the lab of Chris Schell, an urban ecologist who also described himself in his bio as an "Afrofuturist, father, and writer" who "weaves his lived experiences as a Black man and Californian to coproduce justice- and equity-centered research programs." Stocky and bald with a genial dad vibe, wearing khakis and a zip-up Cal fleece, Chris seemed slightly harried that morning. His family had recently moved to the nearby city of Richmond, and he'd already battled Bay Area traffic dropping off his three- and six-year-old kids. The previous night at 9:48 p.m., he'd sent an email to everyone about the fieldwork plan. "This may all go to heck in a hamster ball, but the journey is what counts!" he wrote.

Today, Chris's team was setting up motion-activated wildlife cameras to capture glimpses of mammals such as coyotes, raccoons, foxes, and bobcats. The researchers would split into three teams, each visiting sites in the East Bay such as parks, shorelines, a golf course, a cemetery, and homeowners' backyards. At the lab, several students and postdoctoral researchers trickled in over the next half hour. Rows of boxy devices, which turned out to be the cameras, were stacked on a counter; the students formed an assembly line to slide the cameras into metal frames. Chris started rallying the team.

"We have quite a bit of ground to cover and a small amount of time," he said. Each group needed to pack locks and cables to secure the cameras to trees, along with laminated signs explaining the project. The signs were emblazoned with the motto "urban ecology * conservation * justice" and a black-and-white drawing of a coyote and raccoon facing outward, the raccoon's black markings over its eyes resembling a superhero mask. Chris later told some of the grad students that the lab was in "Marvel phases," a reference to the movie franchise's stages in which the characters advanced through their epic journeys. "We're in phase one right now," he said. "We gotta get to phase four and five."

Over the last couple of decades, wild animals have been spotted in unlikely city landscapes. A coyote wandering into a Quiznos in Chicago. Two raccoons rummaging through an office (and eventually breaking into a tin of almond cookies) in a closed Chase Bank in Redwood City, California. A mountain lion caught on tech billionaire Marc Benioff's home camera, sauntering through the upscale neighborhood of Pacific Heights in San Francisco. And as parts of the globe become increasingly developed, humans and wildlife will live in closer and closer quarters. More than half of humans now inhabit urban areas, and that figure is set to rise to 68 percent in the next few decades. Expanding cities not only infringe on wild habitat but offer appetizing buffets in dumpsters and trash cans for animals to dine on.

For Chris, cities are the ultimate place to explore our relationship with wildlife. "Nowhere on this planet, even in the most remote areas of the planet, are wildlife not experiencing some direct or indirect effect of people," he told me earlier. To conserve biodiversity and live sustainably, we needed to figure out how to coexist with other animals. While we might see urban areas as fortresses of civilization, severed from wildlife, the reality was that we were living side-by-side with critters, from the birds warbling in our yards to the raccoons rooting through our trash bins. "In cities, thinking about conservation has you flip the traditional paradigm of people being outside of nature, and that the only way for nature to thrive is to get rid of people," Chris said. "That narrative we push back on constantly."

And the team was intent on studying the natural world through the lens of environmental justice. That meant going beyond traditional measures used in urban ecology, such as population density or the amount of greenery. The researchers wanted to investigate the effects of other factors such as income level, residents' race and ethnicity, and the legacy of racist housing policies on wildlife. In other words, "how society influences ecology and back," Chris said. For instance, did coyotes spend more time in wealthy neighborhoods with abundant green spaces? Or were they attracted to areas where cities didn't invest as much money in basic services such as trash pickup, leading to more rodents for coyotes to eat?

Some lab members were investigating animal health and behavior in areas with high pollution, a problem that disproportionately affected communities of color and people with lower incomes. One graduate student was dissecting carcasses to look for parasites in the animals' guts, which could be transmitted to pets or people; another was studying whether animals were bolder in more contaminated areas, and thus more prone to conflict with humans; and a third researcher was planning to test the creatures' problem-solving abilities.

As the team finished their preparations, Chris asked an undergraduate student to pick a camera at random and double-check that it had a memory card. "I'm sure that they do," he said calmly, "but my anxiety is high right now." After she confirmed, we headed out of the building and down the staircase to a parking garage, the researchers dragging equipment in plastic crates. I would be riding in Chris's silver Chrysler van with three graduate students who were visiting from other universities to help out with fieldwork. Our first stop was Sunset View Cemetery & Mortuary in El Cerrito, where coyotes, deer, bobcats, and red foxes had been seen. We paused in the lot of a peaked chapel with a fountain and then drove past manicured green hills of gravestones.

After checking out a couple of potential spots to place the camera, the team chose a site in a small enclosure set apart from the graves. They piled out of the van and strapped the lock box containing the camera to a tree trunk, wedging sticks behind the box to angle the lens correctly. Sam Kreling, a PhD student from the University of Washington in Seattle, kneeled in a patch of dirt nearby to pretend she was a coyote and looked at the camera; the "eye" was supposed to point straight into her face. "Sam, back up a couple more steps," Chris said.

"I think right now it's right above my head."

Chris pushed another stick behind the lock box to angle the camera downward. After checking the camera settings, he paused for emphasis and said, "No matter who you are, what your level is, we all keep each other in check. The camera is on. OK? The camera is on. Whoever manipulates the camera should say that audibly. And somebody should, when we leave, say 'The camera is on, did you turn the camera on?' No matter what." Forgetting to turn it on was, apparently, not an uncommon mistake. When you came back and realized you had nothing, he told me later, that was heartbreaking.

After the rest of the group headed back to the van, Chris stayed behind and carefully tucked in the end of a lock box strap and then seemed satisfied. On the drive to the next site, he was in good spirits and started holding forth on the merits of various burger restaurants, proclaiming that as a Californian, "my DNA is methylated with In-N-Out!" We ended up at Point Pinole Regional Shoreline, more than 2,300 acres of meadows, salt marshes, eucalyptus groves, and beaches that jutted into San Pablo Bay. After walking down an asphalt path, the team realized that their intended site was blocked by a marshy area, and it would be easier to drive there instead. But on the way back to the parking lot, someone spotted a trail likely made by wildlife: a long, dark indentation leading into the grass.

"Follow it," Chris said decisively. "Let's see where it goes." We traipsed through the meadow; when we came across the remains of a bird, he got excited. "Something died there," he said, "and this is a *very* clear trail." I was surprised at how straight it was, but Chris told me that coyotes will pace the same path over and over again. "That bird that was kind of torn apart—something killed and ate that thing right there," said Tali Caspi, a PhD student from the University of California, Davis. "And that something was probably a coyote."

In some urban areas, encounters with wildlife are nothing new. For instance, coyotes have been common in Los Angeles throughout the city's history. But in other parts of the United States, humans and other animals are brushing up against each other more often than they used to. In the Columbus, Ohio, area, sixty years ago, "we literally didn't have coyotes anywhere, not rural or city," said Stan Gehrt, a wildlife ecologist at The Ohio State University in Columbus. "They weren't here."

Then the animals started appearing more frequently in big cities east of the Mississippi. Scientists could only speculate why, but coyotes might have thrived partly because wolves and other predators were wiped out and forests were turned into farms. In the Chicago area, where coyotes had been only rarely sighted before, "it came out of nowhere," Gehrt said. They ended up near Navy Pier and on the runway of O'Hare International Airport. In San Francisco, coyotes had been driven out of the city by the 1920s but started moving back in the 2000s; they were spotted in the Presidio, a stretch of wild land near the Golden Gate Bridge, and they quickly occupied other city parks.

Flourishing animal populations in cities, along with increasing urban sprawl, set the stage for what scientists benignly call "human-wildlife conflict." This is an old story that has been unfolding "since *Homo habilis* stood up," said Gabi Fleury, a carnivore biologist at the University of Wisconsin–Madison. Hunter-gatherers had to ward off large predators; scientists have found skulls of our early ancestors with punctures, which might have been bite marks made by a saber-toothed cat, hyena, or leopard. And wild carnivores have been attacking livestock "since the first person put a bell around the neck of a goat and set up shop," Fleury said.

In some rural parts of the world, the clashes are terrifying and destructive. For instance, elephants raid crops in Asia and sometimes trample people when startled in the forest or when farmers try to scare or kill the animals. In cities, conflicts could encompass anything from a wild boar digging up a garden to a car crashing into a deer. Coyotes in North America sometimes kill cats and small dogs. Attacks on people are rare, though a few coyotes in San Francisco, likely emboldened after being fed by humans, acted aggressively toward kids and were shot by officials.

Encounters with wild animals might feel like an affront to city-dwellers who see the creatures as intruders. But "it's just not that they have come to the cities," said Sayantani Basak, a wildlife ecologist at Jagiellonian University in Kraków, Poland. Humans have encroached upon other animals' land too: "We have come to them." This mindset goes against our typical inclination to control our environment, to dictate what belongs where. We consider ourselves "the highest form of living beings," Basak said. "We want to be dominant everywhere." And continually transporting animals back to wild habitat just isn't practical, she said: If you say that they should go back to the forest, where are those forests, and how much is left?

Instead of killing or moving the animals, some scientists are arguing for coexistence. In the Presidio in San Francisco, for instance, certain trail sections are closed to dogs during pupping season, when coyotes are more aggressive. And urban wildlife can benefit cities too. Coyotes kill outdoor cats that are decimating native songbirds, Chris pointed out. (A cat owner himself, he said that people who wanted their pets to avoid this fate could keep the animals indoors, build a catio, or leash or supervise their cat.)

And coyotes can eat a lot of rats, which might reduce the spread of disease. Chris liked to paraphrase a line from a Batman movie: "They're the heroes that we need," he said, "even though, for some, they're not the heroes that they want."

Coyotes were common in my Washington town, though I'd never seen one myself, and local Facebook group threads encapsulated the apparent long history of disagreement on the subject. Some people told stories of a teacup poodle dashing out the door and immediately getting snatched, or of finding the remains of their cat in their backyard. One person advised others to be careful around coyotes that weren't afraid of humans, and then added the weary disclaimer, "I'm sharing awareness. I'm not starting a 'I hate coyotes' thread." Undeterred, someone replied, "Seriously, the coyote-bashing around here gets so, so old… We built stuff in their living room. No reason to fuss about them, just share the space."

When I'd talked to Chris's team, I asked about the ultimate purpose of their research and what it would mean if, for instance, they found that coyotes were more common in wealthier areas or neighborhoods with more garbage lying around. One lab member explained that residents in that area could be taught how to keep their pets safe, or the city could improve trash pickup services. I didn't feel satisfied with the answer, and I realized it was because I wanted them to tell me definitively that coyotes were either good or bad—that we should save them or get rid of them. But coyotes couldn't be categorized that way; they weren't a doe-eyed endangered species in need of rescuing or a scourge to be eliminated. They just were.

I looked up the superhero quote that Chris had mentioned,

and the exact line, spoken by police commissioner Gordon about Batman, was this: "He's the hero Gotham deserves, but not the one it needs right now." On an online message board, people gave lengthy, detailed interpretations of the meaning: "He's not the pretty, polished hero that humanity longs for," one person wrote. "He's the anti-hero, the outsider, the one who'll do the dirty work." Some people thought the quote ought to be the other way around. "Shouldn't Gordon have said, 'He's the hero Gotham NEEDS, but not the one it DESERVES right now,'" someone asked. "No," responded another. "What Gotham deserves is a terrifying, nightmarish being; one that frightens not just the criminals he fights, but the people he protects."

The disagreement over the quote seemed to mirror our ambiguity about coyotes, their place in our world as a figure of fascination and fear, our opposing desires to be connected to nature, with its unpredictability and moments of darkness, and to be separate from it, ensconced in our castles. What *do* we want? What *do* we need? What *do* we deserve?

Burke Museum, Seattle, Washington

In the lab, four coyote carcasses lay on a table. I felt bad for them; they were curled up, paws tucked in, one with a viscous, dark red pool of blood under it. One animal looked surprisingly alive, the air from a fan ruffling its fur and its face caught mid-snarl. "He hasn't fully defrosted either, which makes me a little bit nervous," said Yasmine Hentati, a PhD student at the University of Washington.

We were at the Burke Museum, a building on campus with reconstructed dinosaur skeletons, lofty ceilings, and glassed-in labs visible to visitors. The lab where Yasmine worked was across from a forest ecosystem exhibit and filled with odds and ends such as detached bird wings, rows of creepy eyeless rodents, and a Komodo dragon skull. That day, her team was going to dissect several coyote carcasses. When I met Yasmine, she was wearing a Burke Museum T-shirt with silhouettes of various creatures printed on it. During one of my later lab visits, she pointed out to Jeff Bradley, the museum's mammalogy collections manager, that there was no coyote on the shirt, and Jeff said, "No one likes coyotes. Well, I mean, the majority of people. That's what I mean by nobody."

"I love coyotes," said Yasmine and another student simultaneously.

"You're all correlated data points," Jeff said. "You're all in a coyote lab."

Yasmine had been Chris's first graduate student when he worked in Washington and had helped set up camera traps in Tacoma, about thirty miles south of Seattle. Before that, she'd had a series of wildlife-related jobs: studying deer and ticks in Maryland, tracking coyotes at night in the Chicago area, trapping bobcats by catching their attention with broken CD fragments and feathers. Yasmine was now investigating the effects of pollution and other factors on urban animals, while Chris co-advised her project remotely from Berkeley. She was looking for signs of disease: Heavy metals such as lead could dampen an animal's immune system, which might make it more vulnerable to parasites. Her "shoot for the moon goal," she told me, was to dissect two hundred coyotes and two hundred raccoons. The carcasses would come from areas with varying levels of pollution, including city neighborhoods, suburbs, and exurbs; she'd been picking up roadkill and animals that had been killed because government officials had deemed them unsafe after conflicts with people, as well as animals that had died or had to be euthanized at wildlife rehabilitation centers.

On one of our calls, Yasmine had told me that the dissection would be a "bloody scene resulting in a lot of plastic bags." At the lab, I asked if she'd gotten queasy when she started the project, and she said that during her childhood, her family had celebrated the Muslim holiday Eid by slaughtering a sheep. So this wasn't new to her; mammals were similar inside.

The activity started quickly as team members arrived at the lab, everyone pulling on gloves. Erica Leiserowitz, a project volunteer, cleaned a bloody tray in the sink; Laura Prugh, Yasmine's co-adviser and a wildlife ecologist at the university, read ID numbers for the coyotes out loud; Yasmine gave the team instructions for taking whisker and hair samples. Near the door, an ominous-looking set of knives of varying sizes was mounted on a magnetic strip, next to a framed certificate of a rainbow-maned unicorn. Yasmine offered me a lab coat.

"Will there be...splatter?" I asked.

"If you're getting really close, there might be."

I hung my fleece on a chair at the edge of the lab, and Laura told me I might want to move it. Apparently, the splatter could extend farther than I thought. I moved my fleece.

Yasmine and Laura measured two coyotes and then started dissecting one of them while Erica and an undergraduate volunteer worked on another carcass. Yasmine went over to the knife display. "Laura, do you like a larger knife or a smaller knife?"

"I'm mainly used to scalpels." Tool in hand, Laura carved the body, opened up a bloody cavity, and inspected the state of the lungs, heart, and liver. A distinct meaty smell started to permeate the air. They placed the liver and gallbladder samples in a square plastic container.

"Where was that lymph node?" Yasmine asked.

"Here," Laura said, directing her to the left armpit.

"Can you grab that blood clot?" Laura picked up a dark slimy thing and handed it over.

The meaty smell was getting stronger, warm and ripe. Not a putrid rotting odor, but not a delicious barbecue odor either. The team seemed unperturbed. Yasmine and Laura glopped the intestines into a large Ziploc, and Laura methodically sliced through the kidney like a sushi chef.

Then Laura started sawing off the head. This step was not the most appealing part of the job, Yasmine said, but they needed to skin, clean, and save the skull for the museum collection. Yasmine held the coyote's body while Laura twisted off the head with a cracking sound, and my stomach turned.

The grossest part was the eyes, Laura said: "It reminds me of boba." This elicited a "Nooo" from others. "Sorry, did I just ruin boba for everybody?" As she and Yasmine worked, they chatted about birthdays coming up and the house that Yasmine had moved into. "I'm going to do a vegetable garden for the first time, so I'm excited," Yasmine said, while Laura cut a piece of flesh from the bottom of the skull. After several more minutes of dissection, Laura said, "Okay, eyeball time?"

I quietly retreated and turned to face the wall.

"I'll pull it out," I heard Yasmine say.

"Oooh. I cut the fluid."

After they finished with the carcass, Yasmine told Erica that intestinal dissections would be happening that afternoon. "Do you want to stay?"

Erica hesitated. "Up to a point?"

"It's not as gross as..." Yasmine paused and then said the new protocol was better. "It's *less* gross now."

After each intestinal dissection, the team would save and inspect any worms under a microscope. Yasmine also planned to look for parasite eggs in scat samples that she had been collecting from parks. Some parasites could be transmitted back to pets or people: Coyotes in parts of North America had been found to carry a type of tapeworm that caused tumor-like lesions in people's livers, which were usually deadly if untreated. Roundworms transmitted by raccoons could kill humans. If Yasmine found that animals in highly contaminated areas had more parasites, those results might add weight to arguments that pollution should be cleaned up or prompt stronger warnings to residents to keep their distance from the creatures. She had seen people hand-feeding raccoons in a park and "that scares me," she said. "It's just not safe."

Later that day, a couple of undergraduates joined the team, and I told one of them that I'd heard the intestinal dissection would be gross. She replied, in a tone of utmost warmth and youthful enthusiasm, "Oh, but there could be tapeworms in there!"

Richmond, California

After setting up camera traps at the cemetery and the shoreline park, where the team had found the remains of a dead bird, we headed to Chris's house in Richmond. "Y'all have to excuse the mess," Chris said, as we walked along the side to the backyard, which was strewn with a slip and slide, a kid's bike, a booster chair, plastic buckets, and a pair of little lemon-yellow plastic sandals. Later, he would tell me about his time working with captive coyotes, observing the animals' parenting behavior; he'd seen the adults teaching their young to be bolder around people, nudging puppies toward a new object or encouraging them to eat in the presence of humans. "You see that evolution happen in real time," he said, over a single generation. Eventually, he wanted to study how coyotes raised their young in the city. "As a parent, for me, that's what I think about a lot."

The team planned to set up a camera in a wooded area, which was separated from Chris's yard by a chain-link fence. "How comfortable are y'all scaling a fence?" he asked. The others immediately responded, "Great" and "I'm fine." Chris assured us that anyone who didn't want to climb the fence was welcome to stay behind, but I was determined to show I could keep up, though I was the oldest by about a decade and terminally out of shape. Tali and Sam went over first, followed by Chase Niesner, a PhD student from the University of California, Los Angeles. Hoping I wouldn't embarrass myself, I hooked my sneaker toe into the chain-link, hauled myself to the top with some effort, and swung my leg over. Now feeling confident, I jumped down the other side without looking and nearly impaled myself on a long stick poking straight out of the grass. "Oooh," someone said in concern.

I made it through the rest of the camera setup without incident; that afternoon, we drove to the houses of residents who had given the team permission to install cameras in their yards. Chris got the most excited about the last house, on the edge of a wildish hilly area where the sounds of birdsong mingled with the occasional siren. Near the tree where the camera would be installed, Tali and Sam became absorbed in examining a few blackish, dried-up pieces of poop on a manhole cover.

"I feel like this is actually dog," Sam said.

"I don't know, there's a lot of hair in there." Earlier, Tali had told me that dog poop tended to be yellow from the corn in kibble and fell apart easily, while coyote poop was often thicker and firmer, containing hair, seeds, fruit pits, or pieces of bone.

"Is there?"

"It's too old to be able to tell."

Sam poked it with a stick. "That looks *so* doggy though."

"It does look doggy..."

Chris decided to point the camera straight at the manhole. "I mean, who knows?" he said. "We may get a couple of coyotes taking a dump right there." After the team secured the cables, he backed up to look at the camera from the mystery poop spot and said, "We don't get a coyote on this camera, I'm gonna be upset."

In 2020, Chris and seven other researchers had published a scientific paper titled "The Ecological and Evolutionary Consequences of Systemic Racism in Urban Environments." In that study, they described how a discriminatory policy called redlining, which the government had started carrying out and promoting in the 1930s, had left scars on city landscapes.

The best-preserved records of redlining were maps produced

by a federal agency called the Home Owners' Loan Corporation (HOLC), which was created to refinance mortgages. The agency categorized neighborhoods as A, B, C, or D based on factors such as racial demographics and restrictions, the age of homes, and proximity to industrial sites. "A" neighborhoods were occupied largely by white, non-immigrant, wealthier people; they tended to have newer houses, green space, and deed clauses that forbade selling to non-white buyers. "D" neighborhoods had older houses and were often close to pollution hazards such as railroad tracks and sewage disposal facilities; they typically housed certain types of immigrants who were considered "undesirable," Black people, or white residents with lower incomes.

For instance, in HOLC's written reports about Seattle, "A" neighborhoods were described as having "exceptionally well maintained" homes and "wonderful" views of the water and mountains; these high-end areas were also "protected" by "racial restrictions." "B" neighborhoods, while not quite as enticing, were still considered acceptable, partly because they were "practically 100% American" or housed the Scandinavian community. While some neighborhoods were rated "D" for reasons such as transportation problems, lack of schools, or being located near a gas plant, one justification simply read, "This is the Negro area of Seattle." Similarly, in the San Francisco Bay Area, "D" neighborhoods were described as having blocks "scattered with Orientals and colored" or "a highly congested population consisting of Japanese, Russians, Mexicans, Negroes, etc. having a very low income level." In Oakland, a city next to Berkeley, a field agent wrote of one area that "Unless one knows about the colored families living in the district, there is no means of distinguishing their homes from those of their white neighbors. The homes of the Negroes are in many instances better kept than adjoining homes of white owners." Nevertheless, the area was graded "D."

Even "good" neighborhoods could be demoted if people of certain races or ethnicities were deemed to be dangerously close to its boundaries. One San Francisco area was marked "C" partly due to "a distinct threat of infiltration of Negroes and Japs" from nearby blocks. Reading the appraisers' comments, "the racial logic of it all is very plain," said Jacob Faber, a sociologist at New York University in New York City.

What was the purpose of these maps? HOLC wanted a methodical way to determine which homebuyers were most likely to repay their mortgages; in the agency's view, higher rated areas presented a lower risk to lenders, while lower rated areas were higher risk. This assumption was "completely off-base," said Todd Michney, a historian at Georgia Tech in Atlanta. HOLC's own data showed no evidence of such a link; in some cities, "D"-rated areas actually had the best repayment records. The Federal Housing Administration (FHA) also produced its own redlining maps based on some of the same criteria and denied loan insurance to low-rated neighborhoods. Government officials encouraged and instructed real estate professionals and banks to use similar methods, advising lenders to give loans or favorable terms only to buyers in highly rated areas.

While Black buyers had already faced rampant discrimination from private industry in their efforts to purchase homes, HOLC and FHA gave these practices an official stamp of approval. The agencies "didn't invent racism in housing," Faber said. But "they institutionalized it and scaled it up and put the weight of the federal government behind it."

The effect of redlining on a family's path in life was stark. A white family could more easily secure a favorable loan for a new, affordable house in the suburbs built with FHA backing, accumulating wealth through home value appreciation for future generations. A Black family would have been blocked by private industry from buying in the same neighborhood, a practice condoned by the government; even if they managed to purchase a house in a low-rated neighborhood, their loan typically had a higher interest rate or exploitative terms, and their home's value would likely appreciate far less over time. The wealth gap consequences spiraled from there, affecting whether they could send a child to college or could start a business. "I think it's impossible to overstate the impact that these policies had," Faber said.

In their 2020 study, Chris's team found that neighborhoods rated "D" on the HOLC maps still had less greenery, such as tree canopy, than "A" neighborhoods did, raising questions about how redlining continued to influence the flora and fauna in those areas. Before that paper came out, other researchers had reported links between redlining and people's health—for instance, suggesting that residents of "D" areas in California had higher rates of asthma-related emergency room visits. "What Chris did was say, 'you know what, this is relevant for urban ecology,'" said Rachel Morello-Frosch, an environmental health scientist and one of Chris's colleagues at Berkeley. "It was one of those 'Aha' and also 'Duh' moments"—of course redlining and other structural racism would affect wildlife in cities. Those animals could, in turn, influence human health by carrying diseases that were transmitted to people.

And yet people often assume that such policies no longer reverberate through our cities. "There are many who think of race and racism as an historical event," said Maria Miriti, a plant population ecologist at The Ohio State University in Columbus, who has written about the racist underpinnings of the field of natural history. Chris's paper showed that "this isn't something that is in the past," and that "our human choices, our societal practices, are not divorced from the functioning of the natural world."

Cesar Estien, a PhD student in Chris's lab, had started exploring those connections. I met Cesar during my visit to Berkeley; he had an air of quiet, steady diligence, which sometimes manifested as worry that he'd messed up a camera trap setup, and he expressed a penchant for unwinding after work with eucalyptus and moss-scented candles and his Nintendo Switch. Many of his family members hadn't gone to college, and they thought of wildlife as something far away: lions, tigers, whales. "I never grew up as an ecology boy," he said. Still, during his childhood in Florida, he'd puttered in his grandmother's garden, helping to grow sugarcane and banana plants; and during trips to his extended family in the U.S. Virgin Islands, he'd fished in a lagoon. As a Black and Latino kid from a family with lower income who had left their culture to move to the mainland United States—a "pseudo-immigrant" experience, as he put it—Cesar faced certain expectations. His default options were "the holy trinity of Brown families: the doctor, the engineer, or the lawyer," he said in a seminar about his career path. Cesar opted for civil engineering and struggled in the math classes.

Then one day he came across "wildlife biologist" on a list of possible jobs you could get with a biology degree. He'd never heard of the term before. "I just knew it sounded dope and that I could get paid to follow animals and study and learn more about them," he said. Over the next few years, he studied everything from brown-headed nuthatches to purple sea urchins to North American red squirrels. The squirrel project, which took place in the Yukon in Canada, was a turning point for him, like a "plug hitting the socket." In a fieldwork photo, he was holding squirrels curled in his hands, one bushy tail dangling below, with the kind of megawatt smile one would expect from a parent cradling their newborn.

In one study, Cesar examined data from iNaturalist, a participatory science app that allowed people to enter observations of animals, for four cities in California. In greenlined neighborhoods—those with the most desirable "A" designation—users had detected unique species more quickly than in redlined areas. And when he controlled for factors such as neighborhood size and urban development, he found that redlined neighborhoods had far fewer species. Previous research had suggested that redlined areas tended to be hotter, more polluted, and less green, and these environmental hazards "are shaping the type of animals that can exist there," Cesar said.

When I visited Berkeley, Cesar was investigating whether animals in highly polluted urban areas acted bolder. If they did, the creatures' behavior could lead to more conflicts with people, such as rifling through garbage. And if Yasmine also found that animals in polluted neighborhoods had poorer health, her results might connect to Cesar's findings. For instance, perhaps higher parasite levels or changes in the immune system caused the altered behavior.

Through their research, Chris's team was making justice a more legitimate part of ecological research, Miriti said: "The ability to assert that as valid scientific inquiry is important." In another 2020 paper, Chris and other researchers had described the fictional world of *Black Panther*'s Wakanda, where "the intellectual, cultural, social, and scientific contributions of Black scholars are celebrated," and he urged scientific institutions to take stronger action to dismantle racism in academia to "fully realize the beauty and power of Wakanda in our own universe." That article was important, Miriti said, to voice the experiences of Black scientists in spaces historically constructed for white men. "The relationship between identity and inquiry is something that, particularly in science, is overlooked," she said.

Cesar told me that his identity absolutely played a role in his approach to research. Community was a big part of his culture and upbringing—a constellation of relatives helping his mom, picking him up from school, driving him to soccer practice—and now, he relied on residents in the neighborhoods he studied to learn about sightings of coyotes and raccoons. Traditionally, people thought of scientists as sitting in a tower and working out a theory on a blackboard, but urban ecology "has to be done with people and with community because you're in their backyards," he said. "You're on their street." Near the end of Cesar's seminar, Tyus Williams, another Black graduate student in the lab, issued his heartfelt support. "I wanted to commend you as your friend and as someone who acknowledges you as an intellectual peer," Tyus told Cesar. "You're going very far, and your ancestors would be very proud of you."

Burke Museum, Seattle

After the morning with coyote carcasses, I spent my lunch break outside, inhaling the gloriously fresh air. When I returned to the lab, the team had started the intestinal dissections. Riley Messinger, an undergraduate student, squeezed the contents of a coyote's guts into a bowl like toothpaste from a tube.

"Oh lovely," Riley said, and then announced a few minutes later, "I think we found some worms." She snipped smoothly through another segment of intestine. "It's like cutting wrapping paper when the scissors just go right through," she said with satisfaction. I peeked into the bowl and saw a pile of old defrosted coyote feces tangled with long, translucent worms. "Do you want anything with the stomach contents?" Riley asked Yasmine.

"Yeah, because apparently there's stomach parasites too," Yasmine said. She suggested opening the stomach on a tray, in case a whole mouse was inside.

It turned out that the stomach contained about a dozen baby rats. Excitement erupted in the lab. Jeff, the collections manager, popped in amid the hubbub. He and I looked at the cluster of dead rats and said at the same time, "Oh…my…god."

Coyotes have flourished in cities partly because they can eat a huge range of foods. On a critter cam, scientists once saw a female coyote spend an entire night eating nothing but earthworms. In part of Nova Scotia, Canada, where food options were very limited, coyotes had learned to hunt moose in packs, driving them into snow drifts, over cliffs, and into bogs. And they ate everything in between: goose eggs, snakes, rabbits, turtles, lots of fruit. Sometimes they stalked gophers, waiting for one to pop out of a hole and then leaping onto their prey.

Yasmine explained to me how they passed the intestinal contents through a sieve, and then she advised Riley to push the worms down the side of the bowl, like a chef giving a demonstration on a cooking show. Meanwhile, two other students were dissecting a carcass on another table. Between the slimy red organs, the bowl of dead rats, the mixture of feces and tapeworms, and the pile of discarded intestines, it was all starting to feel like a bit much. Riley banged the sieve of worms against the bowl, and I felt a tiny splatter on my hand. "I think I'm going to rinse off those baby rats," Yasmine said.

Thankfully, the team started cleaning up. One person washed a tray in soapy water. Another scrubbed down the table with Clorox and paper towels. The type A clean freak in me started to feel a little better. *It was all going away!*

When I returned to the lab a couple weeks later, Yasmine's mom was visiting. She was chic and elegant in a knee-length, polka-dotted black dress with flowered sleeves, black tights and ankle boots, and gold bracelets. She spoke to Yasmine in a gentle voice in French; I found out later that Yasmine's parents had immigrated to the States from Tunisia, but her mom had grown up in Paris. The day before, she'd accompanied Yasmine to the freezer containing the coyote carcasses, and she told me that she'd stayed outside just in case her daughter accidentally got stuck inside. "Let's see how long I'm going to last here," she said.

On Zoom later that day, Yasmine gave Chris an update; they were on their fiftieth animal dissection. He looked tired, and when she asked how his week was going, he said, "A lot less sleep than I would prefer." But he perked up when she told him about the coyote stomach full of rats they'd found last week. "Oooh," he said appreciatively.

At some point, Yasmine's mom left to look around the museum. When she returned, she saw the newly detached, skinned coyote head on the table and looked horrified. "How was the museum?" Yasmine asked her. "Are you cold?" Judging from her mother's expression, the lack of a layer was not the problem.

Ellie Reese, a genetics lab manager, inspected the skull. "Were you able to get the brain out?" Yasmine asked.

"No, is there a certain way…?"

"What is the best way to get the brain out?" Yasmine asked Jeff, who had stopped by the lab.

"That's not easy," he said.

As Ellie scooped soft pink blobs out of the skull with a little metal spoon, Yasmine told her mom that the museum had a flesh-eating beetle colony. They'd put the coyote skull in there and the insects would clean it right off. "Then they look like the white ones over there," Yasmine said. "It's cool!" Her mom looked unenthused.

Ellie's pile of pink brain tissue had gotten bigger. "That's a lot," I said.

"That's why they're so smart," Yasmine said.

One question occupying biologists was whether cities were making urban animals more intelligent. Maybe by presenting them with new environments and trying to outwit them, we were driving the advancement of smarter animals with better problem-solving abilities. For instance, if we tried to prevent animals from getting into trash cans, "what we might be doing is actually teaching them how to solve harder and harder challenges," said Sarah Benson-Amram, a cognitive ecologist at the University of British Columbia in Vancouver, Canada. But it was also possible that more innovative, curious animals were the ones that survived in cities, she said. And one could argue that cities, with abundant food and fewer predators, were actually easier to live in than the wild.

A postdoctoral researcher in Chris's lab, Lauren Stanton, was planning to test urban animals' learning and problem-solving abilities by leaving out puzzles and recording their behavior on camera traps. Like Yasmine and Cesar, she'd compare how animals performed in less or more polluted neighborhoods. During my visit to Berkeley, Lauren had shown me a possible design: She was thinking of constructing a device that would automatically release food if an animal pressed the right button, like a vending machine. The team was hoping to attach GPS collars to animals; they could also inject microchips, similar to those used by pet owners, to track individual creatures. Lauren's testing device could then monitor how many times an animal had tried to access the food and present slightly different challenges over time.

At the museum lab, Ellie was almost done with the coyote brain. "I'm going to wash my hands," she said. "I don't like this." Later, she found a pink splotch on her lab coat. "Oh god, I got brain on my coat."

"Do you want a new one?" Yasmine asked.

"Nah."

Olivia Cavalluzzi, an undergraduate student who had gotten slightly disgusted earlier while working on the carcass—the sour smell, the gloppy frozen blood—was now in good spirits. She dumped a huge mass of intestines into a bag that Yasmine held out.

"Yay!" Olivia said.

"Yay!" Yasmine said.

San Francisco, California

We'd found a long, tapered, dark brown coyote poop in the grass. It was the day after Chris's team had set up camera traps in the East Bay; today the researchers were visiting urban parks in San Francisco, where Tali would demonstrate how to find and collect coyote scat for DNA analysis. Using the genetic data, they could figure out what coyotes were eating. Our first stop was Bernal Heights Park, tucked away in a residential neighborhood. We had trekked up a hill of patchy grass, with an urban vista laid out below: a stubby grid of houses, swooping freeway ramps, container ships out on the bay, and the jagged peaks of skyscrapers in the distance. Somewhere in the park was a family of coyotes with as many as six pups.

Tali measured the poop with a white plastic ruler, pulled on a glove, and uncapped a small tube of ethanol. To collect genetic data that would allow them to identify individual coyotes, she explained, the best part of the scat was the outer layer; as the animal excreted, the poop scraped the coyote's epithelial cells. With her gloved hand, she pinched a chunk off the end of the poop and poked it into the tube with an applicator stick. After taking another sample from the middle, she sealed the rest of the scat in a plastic baggie and put it in her backpack, like the least appetizing snack ever. As he watched Tali finish collecting the coyote shit, Chris—who had recently been bogged down in administrative tasks at the university—put his hand over his heart and said, with genuine feeling, "This is healing my soul."

I ended up seeing three coyotes that day at Bernal Hill, though I didn't know if they were all different animals. The first was a female on the hillside below us; I caught a glimpse of her looking straight at us through the grass, and then she vanished behind a bush. A few minutes later, we spotted a coyote trotting at the base of the hill. It lay down next to a metal railing and placidly scratched its head with one leg. Someone in an orange-brimmed baseball cap walked down the sidewalk, and the coyote disappeared behind the railing into the greenery; the person, seemingly oblivious, passed right by. "That person is not aware that they were that close to the coyote," Chris said. "They don't call them ghosts of the plains for no reason." The third sighting came on the drive away from the park—a coyote on the hill across the road from a family playing. I tried to take pictures but caught only a couple grainy shots of a pixelated grayish presence, ears pointing up, in the shadow of a bush.

Because they were both predator and, in some places, prey, coyotes were wily, cautious, and supremely adaptable, said Chase, one of the visiting graduate students. Chase, who described himself as a "multi-species anthropologist," had given me a ride from Berkeley to San Francisco that morning; as we waited in traffic to get through the toll booth on the Bay Bridge, he ruminated on what was behind people's fear of the animals, given that domestic dogs were "astronomically" more likely to harm them or their kids. In Los Angeles, an anti-coyote group called Evict Coyotes was pushing for the animals' population to be culled. "Are they afraid of having to encounter something that's wild or uncanny or mysterious or theoretically uncontrollable?" he wondered. "I feel like they tap into some sort of deep-seated anxiety."

At Chris's lab at Berkeley the next day, Cesar showed me the contraption he was using to test the animals' boldness. Leaning on a desk in a narrow office space were four roughly waist-high wooden sticks connected by twine, which formed an empty frame when you stretched them apart. This was what Cesar called a "novel object": If he set it up at a camera trap site, would animals walk straight through, or would they freeze, approach warily, or inch away? So far, he'd installed one at Blake Garden, and for another pilot study, he'd placed a couple of chairs at Hastings Natural History Reservation.

In one video taken at the nature reserve, a skunk backed away from the chair, staring at it, and ran off. A boar at the same site, by comparison, snuffled around the chair, pushed it with its nose, gnawed on the seat, and flipped it over. So far, at Blake Garden, the raccoons seemed to be the boldest animals, Cesar said. He showed me a video of a plump raccoon walking through the wooden frame, seemingly unconcerned. In another, a coyote paused on the trail, ears up, staring at the new object in its path.

He was always amazed when he captured the animals on video. When I'd asked him earlier what it was like to experience the creatures' lives indirectly, Cesar said, "It's reminding me that animals take time for themselves too, in a way. They have their own space."

Over the next year, the lab's projects progressed in fits and starts. By spring 2023, Cesar had set up the four-post object at about two dozen urban locations; he and a few undergraduate students were going through the videos to code the animals' behavior. He'd heard that coyotes were notorious for acting bold in San Francisco's Golden Gate Park, a relatively uncontaminated site, and wondered if, in certain cases, being close to humans who fed the coyotes was driving boldness more than pollution levels were.

Lauren, the postdoctoral researcher studying cognition, had postponed her ambitious plans to construct a vending machine–like device that presented customized tests; the team was waiting on state approval to capture wildlife, so she couldn't tag and track individual animals. Instead, the researchers had used a simple tube containing a salmon-flavored Sheba Meaty Tender Stick with a rope on one end that had to be pulled to access the food. They'd had to end the testing early when rain deluged the region, raising worries that any new sites they set up would be flooded. Still, they now had thousands of video clips showing coyotes, skunks, opossums, and raccoons trying to open the tubes (and sometimes succeeding). So far, the animals seemed to be performing better on average in less-polluted sites, Lauren told me, but it was too early to say whether this was a real pattern or just the result of differences in how species were spread out.

As for Yasmine, she was still deep in dissections. When I visited her at another lab where the team was identifying and counting parasites in summer 2023, she showed me a collection of jars and plastic deli containers full of worms from the coyote and raccoon carcasses. The coyotes' larger worms looked sort of like rice noodles, while smaller worms were clustered in fluffy looking, yellowish-white clumps; if you shook the jar, the clumps broke into flakes that drifted and swirled like a snow globe. Raccoon roundworms, a few inches long, were pointed at the end and closer to a linguine thickness. Like Cesar and Lauren, Yasmine didn't have a clear answer yet about the differences between animals in more- or less-polluted sites. Many of the carcasses so far had come from the same areas, so she might have to rely on scat samples to answer that question instead.

I'd been hoping for more definitive patterns. It would have been easier to hear a story that was obviously good or bad: that coyotes were bolder and more parasite-ridden in highly contaminated neighborhoods, which meant those communities were at higher risk for conflict and disease, which was bad, but if we cleaned up those areas, then that would be good. Stories as simple as superhero movies, tales of heroism and evil. But I knew that ecology often resisted these kinds of stories. Having more green space in cities, for instance, wasn't uniformly positive; in one of Chris's papers, he'd pointed to a study suggesting that tree cover can increase the risk of Lyme disease, which is transmitted to humans by ticks.

In his talks about the entangled issues of urban wildlife and social justice, Chris liked to quote the character Nick Fury from *Iron Man*: "You've become part of a bigger universe," he'd tell the audience. "You just don't know it yet." I'd never been much of a Marvel fan, but after hearing Chris's frequent references, I started reading about the comics that had inspired the endless series of films. Initially, I learned, characters had largely separate storylines, but the creators eventually decided to merge them into one world. "The actions of each had repercussions on the others," wrote Sean Howe in his book *Marvel Comics: The Untold Story*, and each narrative became a thread of a "mega-story"; as one critic described it, the comics became "a larger, forever-expanding narrative." It would be hard to come up with a better analogy for the story that we were all part of, in which humans played a minor role among a wild cast of coyotes, raccoons, skunks, opossums, and parasites, against a backdrop of asphalt, traffic lights, sirens, and skyscrapers; each subplot driving the narrative momentum of the next; each character, large and small, taking up space in a shared world.

Prototype: Emerald Tutu

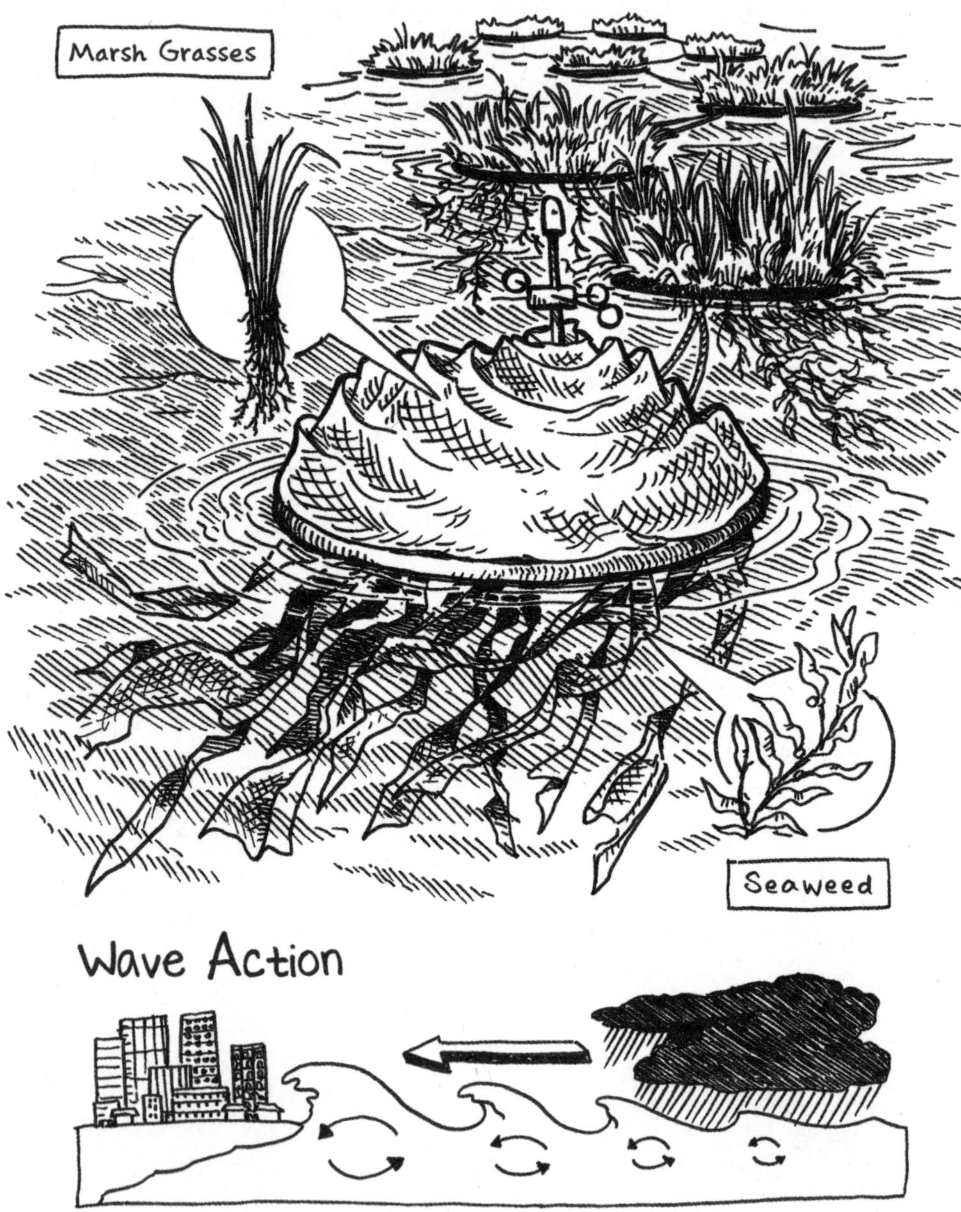

CLIMATE CHANGE

Emerald Dreams

"I don't think we're completely screwed yet."

It was a bit of an odd scene. I was in a wave research laboratory at Oregon State University in Corvallis, standing at the edge of a long swimming pool–like tank of water. To my left was a narrow metal bridge, where Julia Hopkins, a coastal engineer visiting from Northeastern University in Boston, watched the scene intently. To my right was a wall of metal paddles that pushed the water rhythmically to create waves; the pulsing sounded like a giant heartbeat, as if we were inside the body of a whale.

Bobbing on the surface of the turbulent water were what Julia's team called "Emerald Tutus": a flotilla of five-foot-wide circular canvas mats, stuffed with wood chips, stitched with dangling green tulle, and topped with lemon-yellow plastic frills. Jauntily peaked, they looked like giant floating meringue pies.

The researchers, who had constructed about forty mats, had a modest goal in mind. When I had asked them earlier what problem they were trying to solve, Julia said, "Broadly speaking, climate change."

"Oh," I said. "Easy."

The tutus in the tank were artificial prototypes for plant-covered mats, which would be overgrown with marsh grass and seaweed. (In the lab, the yellow plastic material stood in for grass and the green tulle for seaweed.) Hundreds of the tutus could be strung together into an enormous network and anchored to the sea floor. Set afloat in a city's harbor, the team reasoned, these lush miniature islands could dissipate wave energy before water reached the shore. Softening the blow of incoming waves might reduce the flooding expected with sea-level rise and stronger storms.

The vision was lofty; the current reality, not so much. Some of the tutus' seams were now ripping open and leaking debris into the pool. The researchers convened to figure out what to do. "It's not hugely open," said Gabriel Cira, the project lead and an independent architect in Massachusetts, who was inspecting a video he had taken of the mats on his phone. "Just squirting."

One option, Julia said, was to stuff the tutus with fabric and resume the waves. "We have three significant breaks that I can see," she said. "The question is, Do we think stuffing them is enough of a spot fix to keep running? Or—"

"Doing nothing and just keeping running—" Gabe said.

"We cannot do nothing," she said.

"No? Okay."

"No, we cannot do nothing. These will not hold."

They seemed to be at an impasse. The team fell into a deep discussion of tutu technicalities. Tim Maddux, a research associate at the lab, interrupted over walkie-talkie to ask if he could start the next wave test. "Ready to go?" he asked.

Julia looked alarmed. "No!"

The Emerald Tutu project had started in 2017 when a competition at MIT, part of an event series called Climate Changed, challenged participants to devise solutions to hazards such as sea-level rise and extreme storms in the Boston area. Gabe, who was a few years out of graduate school at the time, had previously worked at an architecture firm in Paris called R&Sie(n)—a French play on the word "heresy"—with a reputation for provocative designs and a focus on "strange symbioses between types of beings," as Gabe described it. For one project, he had helped design a house enveloped by hydroponic ferns, which camouflaged the underlying human-made structure. He was drawn to the competition because it was an opportunity to create something "quote unquote visionary," in his words, but also realistic. In college, he had learned about radical architecture proposals that presented amazing visions of future cities—and had never been implemented. He wanted to design something that would actually get built.

Gabe's fascination with urban sites and buildings dated back to his childhood on Cape Cod, where he'd messed around in junkyards and marine yards full of stuff: giant pieces of steel, stacks of wood, cranes, bikes, hunks of cars. He knocked over piles, constructed an enclosure out of pallets, engaged in a haphazard collaging of materials. As a kid, he'd also spent time on the waterfront, sailing and harvesting clams. And he had seen how fragile the shoreline community could be. In 1991, when he was five years old, Hurricane Bob had blasted the region: Power went out, roads were blocked, hornets and yellow jackets swarmed out of destroyed nests.

In the past, cities had typically staved off flooding with a variety of hard, or "gray," structures. Seawalls of stone. Enormous gates that blocked storm surge. Levees overlaid with concrete. Sloped barriers with mysterious names like revetment and riprap.

But these solutions were, "by nature, very static," Julia said. They were built to withstand a certain set of storm and sea-level rise conditions, neither of which scientists could confidently predict. If forecasts were off, a seawall could catastrophically fail. "I wouldn't want a wall in front of New York City," she said. "We don't know what's going to happen with climate change well enough to really be able to design hard infrastructure."

In recent years, researchers had increasingly turned their attention to nature-based, or "green," infrastructure. Marshes, seagrass, and mangroves—trees and shrubs with exposed roots in intertidal areas—slowed down the water flowing toward shore. Oyster and coral reefs made waves break, dissipating energy. And these defenses strengthened as plant matter became denser or reefs grew taller.

But some urban shorelines were already occupied by a tangle of property owners: homeowners, developers, industrial groups. In those cities, it might not be feasible to restore a marsh in expensive coastal areas. "How do you get people out of the way?" asked Edoardo Borgomeo, a water engineer at the University of Cambridge, UK, and an adviser to the Emerald Tutu project. Imagine, he said, telling rich people in Manhattan: "Listen, we're going to move you somewhere in New Jersey." It wouldn't go over well.

So for the competition, Gabe started thinking about the

possibility of a "floating, biomass-based, modular" solution that would be installed out on the water instead. The name Emerald Tutu was a play on the Emerald Necklace, a set of parks developed in Boston in the late 1800s, that suggested a sense of humor and a "sassy queer attitude," he said. (This attitude would later manifest itself in a video of a drag queen with a pink bob and shimmery green nails gamely explaining the project's esoteric scientific details—such as the tutus' "local mesh network dynamics" and wave interference "similar to a Helmholtz chamber"—while also bemoaning the "fashion disaster" in the city's Beacon Hill neighborhood.)

Once he had the basic concept, Gabe realized that he needed scientists. Friends put him in touch with Julia, who was working in the Netherlands at the time. To counteract dune erosion and prevent flooding, that famously low-lying country relies heavily on a strategy called nourishment, which means dredging millions of cubic meters of sand every year from the North Sea and dumping it onto coastlines. Julia's job was to improve the process of piling sand underwater near beaches. The research was very Dutch-centric, "which is perfectly fine," she said. "But my imagination is captured by things New York City can use to hold back the water." That, she said, "is the biggest challenge you could possibly think about." She was immediately taken with the tutu idea. In addition to avoiding the legal morass of regulations and space constraints that often bogged down projects on shore, the tutus were more aesthetically pleasing and environmentally friendly than a sea wall. And the mats could be made out of relatively cheap materials. "I was like, yeah, this actually stands a chance at being put in the water," she said.

The modular design also meant that the tutus could be

adapted to different situations; as the effects of climate change became clearer, mats could be added or taken away. That flexibility was critical, Borgomeo said. "We want to invest in things that can be changed as the future unfolds," he said.

After Gabe recruited a few more researchers, the team decided that the mats should be made of flexible material, with marsh grass planted on top. Seaweed would likely colonize the bottom of the mats on its own; these algae often anchored themselves to the sea floor but could also attach to other surfaces. In spring 2021, the team assembled a prototype and tied it to an abandoned pier in Boston Harbor: a net-enclosed burlap mat filled with wood chips, stringy coconut fiber, and foam for buoyancy.

Now Gabe and Julia—along with Jena Tegeler, a landscape designer then based in Cambridge, Massachusetts, and Tyler McCormack, a PhD student in Julia's lab—had set up shop at the O. H. Hinsdale Wave Research Laboratory at Oregon State University to construct and test smaller artificial mats. The team wanted to measure how much the tutus reduced wave energy when pummeled with waves of varying heights and frequencies. To figure out how much protection the seaweed added, the researchers planned to cut off the hanging tulle partway through the experiments and monitor the difference in performance. And they would test different configurations, with mats packed close together or spaced farther apart.

During the year before my wave lab visit, the forecasts of climate change disaster, which had seemed so abstract, felt more and more real: wildfires in California, apocalyptic smoky skies in the beauty-soaked city of San Francisco where I'd spent my early twenties, and the staggering heat wave in the Pacific Northwest that had roasted the interior of my house, even in the shade of towering evergreens, to more than ninety degrees. At one point, I asked Julia if there was any hope of adequately addressing climate change. "Can we actually problem-solve our way out of this?" I said.

"We totally can," she replied with confidence.

I was surprised by her optimism. "You really think so?"

"I think we can," she said. She explained that we already had a way forward; we needed to restore marshland, forests, the ecosystems that naturally removed carbon dioxide. "I don't think we're completely screwed yet."

"You don't think we're too late?"

She said, "I think if we start thinking that we're too late, then we have lost."

The wave lab was a low, boxy building on a grassy expanse outside the bustle of campus in Corvallis, a pleasant college town hugged by the Willamette River and forested wilderness. When I visited the team, I found them working in the directional wave basin, a space about the size of a high school gymnasium. About two-thirds of the space was taken up by the tank; the other third was strewn with various equipment, including a yellow boat replica, orange pylons, a bucket of wood chips, shovels, coils of rope, and a large metal booth. On the floor in front of the tank were foam mats made of brightly colored, interlocking squares, the kind used in preschools.

In the weeks before my visit, I had watched the tutus take shape on one of the wave lab's live-streaming webcams. Several people in orange construction vests—whom I assumed were Gabe, Julia, Jena, and Tyler—had been busily occupied. One morning, one person had unrolled white canvas on the floor and crawled around on hands and knees to cut out a circle about the size of a living room rug. Another person picked up the canvas circle, which resembled a huge tortilla, and slung it onto a pile of others. Someone else was hunched over a sewing machine, with unruly bunches of tulle nearby. Slowly, over the next week, completed tutus started to appear, decorated with mesh on top and tulle skirts peeking out underneath.

When I arrived at the lab, forty-four tutus were floating in the tank, and the team was watching from the bridge. They were running some tests but were still in "baby wave territory," as Julia put it. She had a decisive manner, upright posture, and a habit of standing with her feet planted far apart as if bracing for impact, partly a holdover from her days as a fencer in high school and college.

Julia had been aware of the climate change crisis since she was twelve years old. She read *Silent Spring* by Rachel Carson and bugged her mom to start recycling; news reports about the weather changing worried her. She decided she needed to do something about it.

At first, she thought she'd be a lawyer fighting for conservation, then a biologist. But after starting college at MIT, she'd started building random things in workshops—rollercoasters, foot bridges, costume props. Julia realized that as an engineer, she could think of a problem and figure out how to solve it. She decided to pursue fluid mechanics because the math behind it was "*really* gorgeous," she told me in a rapturous voice normally reserved for discussing Rachmaninoff concertos or Cézanne paintings. With solid objects, you didn't get a smooth continuum; imagine applying pressure to a concrete structure until it suddenly broke. "That, mathematically, is quite jarring," she said. Fluid would always be an elegant curve. "There's no start and stop," she said. "It all flows into itself." She focused on oceanography, and then—after Hurricane Sandy hit in her first year of graduate school—on the coasts. Though her career path had taken some turns since her twelve-year-old awakening, Julia's goal stayed the same. As a kid, her main thought had been, "I must do my best."

The tutu team was going to start running bigger waves soon, which would definitely test the system, she said: "It's getting us closer and closer to hurricane force." Sensors at the bottom of the tank detected the pressure of water above, which could be converted into essentially a topographical map of wave heights, while instruments at the surface recorded the water level. By measuring the wave height behind and in front of the mats, the team could determine how much the tutus were dissipating wave energy.

The water swelled, crested through the tutu network, and crashed on the "beach" on the other end. Some of the tutus had a strange apparatus on top: metal rods sticking out like antennae, capped with small, shiny gray balls. The contraptions gave the tutus the appearance of alien spacecrafts that had landed on the ocean and were beaming a message back to the mothership. Their purpose, the team explained, was to capture the mats' precise movements. Cameras on the sides of the tank shone infrared light on the balls; the researchers could track the reflections and thus the mats' coordinates.

After a few wave tests, the team ran into a problem. The tutus were connected to each other with nylon rope, but some of the splices were coming undone. Jena, outfitted in oversized brown waders, paddled into the water in an inflatable raft and grabbed the edges of the mats to maneuver around them. Gabe, who was wearing fashionably thick-rimmed pale green glasses and cuffed jeans that revealed bright pink socks, watched her progress from the bridge. "What are you finding?" he called.

"I think it's just coming undone completely," she said. She had a deliberate way of speaking, never raising her voice more than a few decibels even when the rest of the team was in spirited debate.

Gabe's theory was that splices performed with a certain motion—folding the rope "left-right-left"—were weak, while those done with a "left-left-left" motion were holding up.

"This one's fine," Jena said, looking at another connection.

"But check if that's a left-right-left or a left-left-left."

"It's a left-right-left."

Julia sounded skeptical of Gabe's diagnosis. "There's no inherent reason why the left-right-left shouldn't necessarily work," she said. As they debated, Jena methodically inspected a connection below the bridge. She floated from tutu to tutu, patiently attending to each one like a fisher repairing her nets.

Cities built hard infrastructure in an era when the attitude was "You're going to conquer," Greg Guannel, a coastal engineer and director of the University of the Virgin Islands' Caribbean Green Technology Center in St. Thomas, said. "Whatever the problem, we're going to fix it." But by fixing a problem, engineers created new ones. Take seawalls, for instance: A wall can worsen water quality by blocking the exchange of water with the ocean. Sometimes it disrupts habitat for fish. It shunts wave energy to the sides, which is what it's supposed to do—but at some point, the wall ends. Then the ground at the sides of the seawall erodes, essentially moving the coastline inward and increasing the risk of flooding in neighboring areas. Those neighbors could put up their own walls, but that results in "death by a thousand seawalls," said Katie Arkema, an earth scientist at Pacific Northwest National Laboratory in Seattle. "Is that what we want our coastline to look like?"

The Netherlands—which is under constant threat from sea-level rise—has its share of hard infrastructure, including storm surge barriers, a dike so huge that it's visible from space, and a gate with Eiffel Tower–sized doors that protect the Rotterdam area. But since 1990, the country has been committed to using "soft" solutions for coastal defenses whenever possible, such as replenishing coastlines with sand, said Renske de Winter, a coastal engineer at Deltares, a research institute in Delft.

Nature-based solutions, proponents argue, offer a more environmentally friendly approach. They provide habitat, clean the water, and sometimes sequester carbon or double as pleasant spaces for recreation and tourism. While not as imposing as a seawall, they can protect coastlines in other ways. Imagine waves rushing toward a marsh, a bed of seagrass, or the elaborate roots of mangroves. The vegetation creates friction, absorbing the water's energy. Oyster and coral reefs also raise the "ground level" of the ocean floor, making it shallower and causing waves to break. And in some areas, plants trap sediment, reducing shoreline erosion. "They're all a little different," Arkema said. "That's part of the beauty," she added. "You have multiple lines of defense."

By providing seaweed dangling from the water's surface, rather than growing upward from the sea floor, the tutus might add a new type of defense. Julia speculated that this arrangement could be advantageous because most of the wave energy was concentrated at the surface. For the artificial prototypes at the lab, though, the team was struggling to mimic hanging seaweed. Using tulle hadn't worked: The fabric was too delicate, and they hadn't added enough of it, so it created almost no resistance to the waves. A week into that decision, the team realized, "This isn't going to do anything," Jena said. The new plan was to use pieces of canvas left over from cutting out the mat circles, which were already in a "perfect stringy shape," Gabe said. They'd lift the tutus with a crane and attach canvas bundles to the tulle.

Toward the end of my first day at the lab, in the midst of the rope connection repairs, Julia started questioning whether they needed to use the crane. They were planning to partly drain the tank; if the new water level wasn't higher than their ribcages, they could just wade into the pool and reach underneath the tutus to attach the canvas.

Tyler disagreed. "When I was swimming around there today, it's very difficult to navigate," he said. "The tulle is, like, sticking to every part of your body."

"Okay," said Julia, abandoning her idea. "We're lifting them up with the crane."

Gabe was unconvinced that the sticky tulle would be an issue. "That's because he was swimming," he said.

"But if you can't navigate it, which means you can't get underneath it without completely getting entangled in tulle, it's actually going to be easier to lift them up."

"So plan is as planned," Gabe said. "Plan remains as planned."

"Until it changes," Tim said.

Julia said, "Jesus, right?"

The plan changed the next day. Things were going awry; this was the point at which the seams holding the canvas pieces together ripped and started leaking debris. It wasn't the first time the tutus had split. They'd used a slightly different design than the Boston Harbor prototype to fabricate many tutus quickly, which had caused too much force to be applied to the seams. And during tutu construction, the team's industrial sewing machine had broken; they had to stitch some remaining seams with nylon rope and a handmade "needle" fashioned from a bolt. Those weaker hand-sewn segments were now coming undone. "It's a series of little decisions that have kind of added up," Jena said.

The team paused the wave tests and turned their attention back to the seaweed question; they weren't sure if they had enough canvas to properly mimic lush vegetation. Julia, who had been scrolling through fabric options on the JOANN website on her phone earlier, was talking animatedly to Gabe on the bridge about a possible solution as Jena and I walked up. "We wouldn't have to lift them up because we wouldn't have to reach underneath them," she was saying. "What if—"

"I'm really tempted to flip them," Jena said, in an almost imperceptible voice.

"—we took a play mat and put a bunch of stuff on it, like bought a bunch of felt and just stapled it on or something?" Julia said. She was talking about the square foam play mats that were lying around the lab. They could slide the foam, decorated with dangling fabric strips, under the tutu and secure it to the bottom. "We might just be able to get a denser canopy that way," she said.

Gabe was thinking about the amount of work involved in attaching the canvas strips. "That's like forty-four little underwear guys…" He was referring to the fact that the leftover triangular canvas pieces bore an unfortunate resemblance to undies.

"I mean, we're going to have to mass-produce whatever we're doing no matter what," Julia pointed out.

"That's why I'm wondering about flipping being the answer," Jena said, very slightly louder this time. If they rolled the tutus over, the yellow plastic frills could become the simulated seaweed on the bottom.

"The problem with flipping is that the yellow stuff isn't long enough," Julia said. "I don't think it would do much different."

"You might be right. It's basically just putting a fuzzy bottom on it."

"That is actually not nothing," Julia conceded.

They started brainstorming ways to add more material and decided to zip-tie the canvas bundles to the yellow frills. "Then we just flip," Jena said. "Which will be fun." She looked pleased.

The three of them moved to a small office cluttered with tutu construction materials, and Julia and Jena settled into wooden wheelie chairs while Gabe folded a canvas fragment into an upside-down flower shape with dangling petals.

"Actually, if we just do that…" Julia said.

"That's actually the perfect—"

Julia approved. "That's great. Wonderful. Done. Okay. Fantastic."

They decided to place a foam nub in the center of each canvas triangle, fold the material over it, and zip-tie the top like a collar to hold the foam in place. "Great," Gabe said. "We're making decisions." Energized, the team moved to the main part of the lab to start production. They sat down cross-legged on the play mats, like a group of kindergarteners at crafting time, and started assembling the canvas bundles; they would need to make nearly three hundred. As they finished each bundle, they tossed it into a growing pile. A peaceful, meditative mood set in.

I asked them how real climate change felt in the Boston area, compared to the Pacific Northwest where we'd been battling wildfires and breathing smoky air. Gabe said that Boston hadn't been slammed by a major hurricane for about thirty years. "It just hasn't happened yet, and it will," he said. "That will be a moment of truth."

When I looked up the research on how bad we expected the situation to get, the numbers fell into a familiar drumbeat of doom: More than sixty billion dollars of damage from Hurricane Sandy. Trillions of dollars of destruction from coastal flooding caused by rising waters worldwide over the next eighty years unless we took urgent steps.

The flooding risks fell into a few buckets, so to speak. First, there was the storm surge, a pile of water pushed toward shore. On top of the surge, there were "huge waves bamming," as Julia put it. If you were unlucky, all of this happened during high tide. On top of the surge, the waves, and the tide, there was rain. And land was also sinking in some areas, making flooding even worse.

As sea levels crept up, neighborhoods farther and farther inland would be exposed. For instance, a report on Boston's climate vulnerability warned that by the 2070s, severe floods "will expand to vast areas of the city." In the South Boston neighborhood, storms that have a 10 percent chance of happening each year could inundate areas about a mile inland from the coast.

What did this mean for cities, beyond flooded basements? Electricity and public transportation systems could go down.

Hurricane Sandy had temporarily dragged business in a major metropolitan area to a halt. Flooded roads could prevent trucks from transporting essential goods; in waterlogged soil, trees could fall more easily, posing another risk to power lines. After Hurricane Ida, when power outages had knocked out air conditioning in homes, more people in New Orleans died from the heat than from the storm itself.

Exactly how much flooding the tutus could prevent remained hazy. Nick Lutsko, a climate scientist at the Scripps Institution of Oceanography in La Jolla, California, who was collaborating with Julia and Gabe, was circumspect. Because the tutus were porous, he said, they wouldn't block all the water from storm surge. To protect cities, other solutions would probably need to be deployed too. "It's not like you throw a bunch of floats in the water and then your coastline is safe," he said.

During my lab visit, though, Julia envisioned a scenario in which the tutus might reduce the surge. We were sitting in the lobby, a long room decorated with a painting of a tsunami wave, stylized curls of blue, green, and gold. She speculated that in New York, the tutus could be strategically deployed south of the bridge that connected Brooklyn and Staten Island. If the mats had been present during Hurricane Sandy, perhaps they could have shaved a couple feet off the surge entering Manhattan. A couple feet would have prevented a lot of flooding. That, she said, "would have not been nothing."

After the team finished making a pile of canvas bundles, I waded into the tank. The water was about knee-high and murky, with wood chips floating on the surface. Up close, I could see that the yellow plastic material on top of each tutu was arranged in a spiral, reminding me of an elaborate layer cake—though with its steel-wool texture and clogged bits of debris, it didn't look particularly appetizing. The mats had a sour, funky smell, like a wet bathing suit left in a car too long.

Gabe and Jena, outfitted in waders, had piled the canvas bundles into the raft and were attaching them to the mats, where they stood upright like stiff bouquets. After I zip-tied one bundle to a tutu (and Gabe generously pronounced my work "beautiful"), I returned to the edge of the tank to collect more. Water sloshed around me, and slimy tulle brushed against my legs. "Welcome to the swamp," Gabe said, when a couple of undergraduate students showed up to help. For the rest of the afternoon and the next morning, the team made their way methodically through the canvas bundles, bent over the tutus like industrious farmers planting their crops.

As the team worked, Tim showed me the wavemaker machinery at the far end of the pool. A row of motor-driven triangular bars pushed the metal paddles back and forth to create surges of water. Behind the machinery were tall cabinets holding electronics; they were the brain of the operation, taking commands from the control center and translating them into mechanical actions. When we returned to the side of the tank, the tutus, with their perky petaled additions, now looked like a display of outlandish royal wedding hats. The team would have to flip them, each weighing up to 200 pounds, like giant pancakes.

"What do you think?" Tim said. "Is it going to work?"

That was the fundamental question for "green" solutions. Would they work? Before visiting the wave lab, I'd spoken to Dan Cox, a coastal engineer at Oregon State University, whose team had also studied a nature-based solution at the lab, and asked why it was important to conduct that research. “This could all backfire,” Cox said. “There’s almost a drumbeat of ‘Hey, let’s do this, natural systems are better.’ Okay, but what level of performance are you really getting? When a storm comes in, are you going to be okay?”

In some cases, the answer seemed to be yes. Researchers estimated that coastal wetlands prevented more than $625 million in flood damage across twelve states during Hurricane Sandy and $934 million in the Galveston Bay, Texas, region during Hurricane Ike. And natural systems might offer even stronger defenses than hard infrastructure. In one study of Florida Keys neighborhoods after Hurricane Irma, researchers compared the damage to houses behind mangroves and structures such as bulkheads, which are small walls typically built with wood, steel, or concrete. “Very consistently, we saw that mangroves provided protection,” said study coauthor Tori Tomiczek, a coastal engineer at the U.S. Naval Academy in Annapolis, Maryland. Those plant-protected houses fared better under severe storm conditions than others did.

But data on the effectiveness of some nature-based solutions was still lacking. One team analyzed sixty-nine studies of natural features, such as coral reefs and seagrass beds, and found that they reduced wave heights by an average of 35 percent to 71 percent. Most of the measurements were performed when wave energy was low, cautioned study coauthor Siddharth Narayan, a coastal engineer at East Carolina University in Greenville, North Carolina. It’s difficult to collect data during extreme events, he added, though one recent lab study suggested that willow forests could substantially reduce the heights of large waves.

Nature-based solutions came with other caveats too. Performance might decline in the winter, when plant matter wasn't as lush. Natural systems were limited by climate; you couldn't put coral reefs in the Northeast, for example. At certain sites, wetlands could actually increase flooding because they redirected water around them during a storm surge (though this negative effect was still far outweighed by the overall protection that wetlands provided). And though nature-based solutions were flexible—oyster reefs would grow over time and marshes could accumulate sediment—they might not be able to keep up with sea-level rise.

If sea levels rose too quickly for cities to adapt, it was possible that all solutions, including the Emerald Tutus, would end in retreat. Julia estimated that the tutus would become obsolete after a couple feet of rising waters. We were just putting off that moment, and we'd likely remain in denial for as long as possible. "It will only be considered when, really, there's nothing else to do," said Stephanie Kruel, a senior regulatory and resilience adviser at the civil engineering firm VHB who had given informal feedback to the tutu team. "It's never going to be before disaster happens. It will be in the wake of probably several disasters."

As the researchers worked on the tutus, water gushed from a spigot on the side of the tank. The water level slowly crept up, getting dangerously close to the spiral-bound notebook dangling around my neck. I felt like I was standing in an enactment of an overly obvious metaphor: The sea level was rising, and we were frantically trying to throw together a solution before the water swallowed us.

It was time to flip the tutus. Gabe and Rachel Calabrese, a student, grabbed the edges of a mat and rolled it over like a doughnut in a fryer. The exposed underside looked sad: dirty and speckled with brown spots of who-knows-what, limp strands of tulle lying on top. But Gabe looked happy. He, Rachel, and another student, Alec Martinson, flipped another mat. "This smells terrible," Alec said.

A couple days later, the team was ready to run a series of random waves with different heights and frequencies. This was the closest they would come to mimicking the real ocean conditions the tutus would face. The test would take fifteen minutes, "so settle in," Julia said. "Take a deep breath, pray nothing breaks."

After setting up an underwater camera, they radioed the control room to give the green light. A wave roared through the tank. "There's a big one, holy shit!" Julia said. "It's just fifteen minutes of biting my nails." The tutus seemed to be holding up. Another wave came, and then another. Then a really big one. "Jesus!" Julia covered her head with her hands. "Shit!"

My visit ended, and the tutus survived. After returning home, I asked various experts what they thought of the project. The responses were mixed. Ariana Sutton-Grier, an ecologist at the University of Maryland in College Park who has written about coastal issues, called the idea "super innovative, very creative." Putting the mats offshore was a "huge bonus," she added, given the difficulties of finding space along the boundary of harbors in urban areas. The team wasn't just relying on a natural ecosystem but actively engineering a new solution, Narayan noted. By reducing the force of incoming waves, rather than stopping them at the coast with a seawall, "you really get to the core of the problem," said Evelien Brand, a technical manager at Rijkswaterstaat, executive agency of the Ministry of Infrastructure and Water Management in Utrecht, the Netherlands.

But the team would need to figure out what, exactly, the tutus could protect cities from: A run-of-the-mill storm? An extreme storm? Even if they dissipated only smaller waves, they wouldn't be useless; they could reduce flooding during high tide. "But it is important to be clear about what they can do and what they cannot," Narayan said.

Experts also gave approximately a thousand reasons why the Emerald Tutus might fail. It would be hard to get permits to install them, Kruel said. The mats could interfere with boat navigation. Rich people in waterfront condos might object to having floating mats in view. There was the possibility of unintended consequences: Would the tutus shade out plants growing underneath them or reduce water quality? Also, maintenance: How frequently would they need to be repaired? And Guannel expressed doubts that the project would be cheap. "Let's not fool ourselves," he said. If people are out working on a barge, "that's hundreds of thousands of dollars, no matter what they put in the water."

The tutu team already had ideas for surmounting those obstacles. They were working with the US Army Corps of Engineers and public policy experts to figure out permitting, Julia said. Having more space between mats didn't seem to affect their performance much, so perhaps small boats could maneuver between them in calm weather, or cities could designate separate channels for boats. The researchers planned to monitor whether shading affected organisms on the sea floor. They hoped that the tutus could even be a revenue source for less wealthy communities—for instance, if the installation was considered a carbon offset.

Based on the data collected at the lab, the tutu team concluded that the mats worked well in some situations, with caveats. They reduced wave energy by more than 80 percent if the wave length—that is, the distance between two crests—was less than the width of a single mat. The tutus continued to partially dissipate waves as long as the waves were shorter than the width of the entire network of tutus. If a wave was longer than the mat network, the tutus weren't as effective.

It wasn't clear yet if the tutus could reduce hurricane-force waves. But they could offer protection from shorter waves, which was "not nothing," Julia said. Ever the optimist, she said the results were "really good news." Now the team understood the relationship between wave length, tutu network size, and wave energy dissipation, so they would design networks to be at least as wide as the length of storm waves.

On a trip to Boston in fall 2022, a little more than a year after I'd visited the Oregon wave lab, I stopped by a former industrial site near the city's airport. Gabe had grand plans to create an outdoor lab here: He wanted to build a floating frame to hold prototype tutus in a creek at the edge of the property. The site used to be a storage site for fuel tanks, and Gabe saw a kind of poetry in creating new green solutions from the ruins of oil and gas infrastructure. The property was blocked off with chain-link fence, with a rusted "NO TRESPASSING" sign; through the fence, I could see tufts of grass poking through concrete, a discarded Dunkin' cup, and a crumbling white building graffitied with the message "EAT the RICH." Gabe's team had set up assembly materials here, including a pile of wood fragments, bundles of coconut fiber, and a metal basket hanging by chains from a frame, which the researchers used as a scaffold.

Earlier that day, I'd driven north to the picturesque small city of Beverly, where the team had recently deployed a prototype tutu off the coast. Louiza Wise, a recently graduated ecological engineer who had joined the project, met me there and led me around some school buildings, across a lawn littered with leaves, and toward an outcrop. In the distance, a small mound of dark rocks jutted out of the water. About halfway between us and the mound was the tutu, anchored to the sea floor; it was barely visible, nothing more than a beige cap peeking out of the waves. I had to squint to see it against the white glare on the water. A wave washed over it, and it disappeared and then bobbed back up.

I thought of all the experts' reservations about the tutus' chances of success. I'd asked Narayan how optimistic I should be about the project, given the long list of caveats. He told me that every little bit helped: "There's not going to be a single solution for any of these gigantic issues." For nature-based solutions, "we want to see pilot projects," he said. "It is essential that there are people taking that risk." Engineers, he added, "learn a lot more from our failures than from our successes."

"So even if it fails, then we'll have learned something?"

"Yes," he said with enthusiasm. "Definitely."

An article I'd recently read argued that we should be prepared to try every possible solution to the climate crisis. And I remembered Julia's habit of saying "not nothing." That's what the tutus were: part of the everything and not nothing. If we collectively did enough "not nothings," maybe that would be enough.

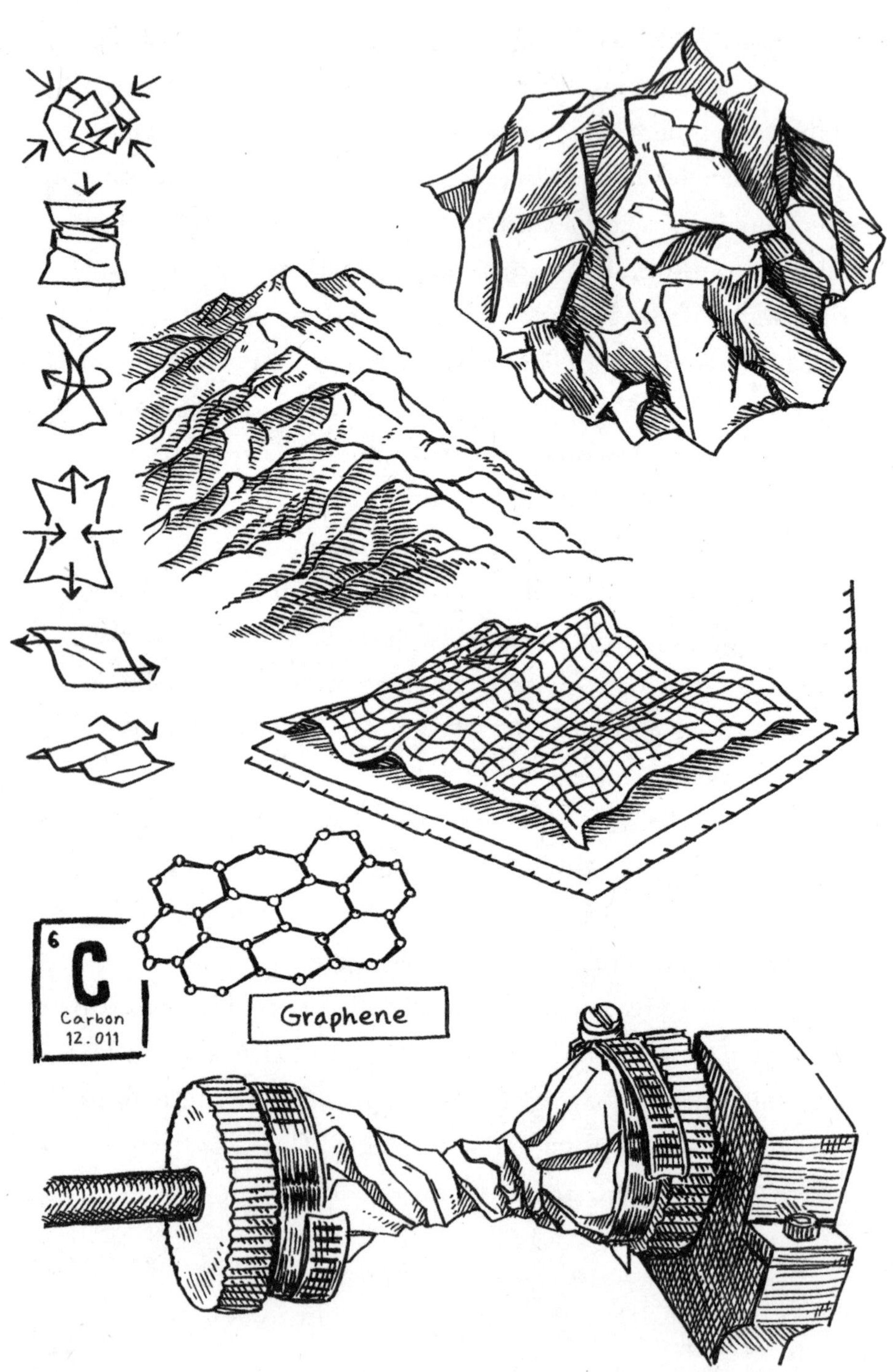
6
C
Carbon
12.011
Graphene

Physics

Paper Labyrinth

"Mathematics, rightly viewed, possesses not only truth, but supreme beauty—a beauty cold and austere, like that of sculpture..."

The physicists were talking, but I had no idea what they were saying.

Madelyn Leembruggen, a physics PhD student at Harvard University in Cambridge, Massachusetts, was on a Zoom meeting with four team members scattered around the country, and she'd just screen-shared a few sets of graphs. "There's a little bit of spread at small *n*, but less higher up," Madelyn said.

Her adviser, Chris Rycroft, drew an arrow to an equation. "I would have thought that this is not really l_c^1, but it would be really just some sort of *delta* l_c," he said. "Some small correction factor."

"So that one would be fitted?" Madelyn asked.

"That's what I was thinking, yeah..."

My bemusement wasn't new; I'd been sitting in on the team's meetings over the last couple of years, and 99 percent of their remarks sailed far over my head. Sometimes everyone would laugh at an esoteric joke or nod vigorously in agreement, having shared a lightbulb moment, and I'd pretend that I understood what had just happened. Still, there was something soothing about listening to them, like hearing people converse enthusiastically in Italian, enjoying the melodious sounds coming out of their mouths, and occasionally understanding a word like *pizza* or *ravioli*.

The topic at hand: paper crumpling. Specifically, what would happen if you twisted a sheet of paper like a candy wrapper or smushed it into a ball? More specifically, if you crumpled it over and over, how would the creases change over time?

Though it might seem impossible to predict anything about the folds that emerged, the team had good reason to believe that understanding was within reach. Several years ago, their collaborators had crushed thin sheets of material repeatedly from the top inside a cylinder and tracked the total length of all the creases. The team found that this number increased according to a tidy equation. In other words, if you added up the crease lengths after crumpling, you could reliably predict how much the total length would grow when you squished the sheet again.

"It is absolutely wild," Madelyn told me. "It feels random to us—it feels like there's no way that could be deterministic."

Madelyn now wanted to know if the same rule held true when the sheet was twisted in the middle or squished from all sides. She'd been simulating the process on the computer, and collaborators at Clark University in Worcester, Massachusetts, had been running experiments in the lab. Chris's hunch was that the pattern was universal. "I've really felt that this basic picture applies more broadly," he said. "I must admit I'd be shocked if it's not there."

The cynics in the audience are now probably sniping, *Shouldn't the greatest minds in the country be working on a more important problem, like solving the energy crisis or probing the origins of the universe?*

But for some researchers, crumpled paper's everyday nature is exactly what makes it worthy of attention. "Part of the role of physics is to explain the world we live in," said Sidney Nagel, a physicist at the University of Chicago, whose team has studied commonplace phenomena such as coffee rings on tables. Whether it's bubbles in fluids, birds flocking in the sky, or fish swimming in schools, "we want to understand what we observe every day," said Martine Ben Amar, a theoretical physicist at the École Normale Supérieure in Paris.

And while paper crumpling might seem pedestrian, it presents a formidable problem. Right now, it's very difficult to predict where creases will appear or how they'll change depending on the way a sheet is crumpled. "Even though we know all these incredible things about subatomic particles and the center of the galaxy and things like that, these everyday phases of matter"—like a crumpled sheet of paper—"hold mysteries we don't really understand," said Chris, an applied mathematician now at the University of Wisconsin–Madison. From that disorder, he said, "emerges a real complexity."

If we could grasp that complexity, physicists might be able to extend the knowledge to other systems. Paper is simply a thin sheet. And in physicists' eyes, "thin" is relative; it means a material with a thickness much smaller than its overall size. The skin of a jellyfish, the hollow body of an airplane, and the outer layer of our planet are also thin sheets. When I asked Nagel if that meant the principles from paper crumpling could be applied to Earth's crust, he said, "That is precisely the hope." For instance, perhaps we could better understand how mountains or hills form when the crust buckles.

And at the microscopic level, graphene—a material made of a single layer of carbon atoms—looks very much like paper when crumpled. Researchers are investigating ways to use crumpled graphene as electrodes, so it's "quite reasonable" to think that paper-crumpling research could help improve battery design, Chris said.

The dream, then, is to find laws that transcend scale. "We are broad brush people," said Narayanan Menon, a physicist at the University of Massachusetts Amherst. Physicists like to explain many phenomena with one idea, like gravity: "The way apples fall is the same way that planets go around the sun."

In the Zoom meeting, the researchers, after sparring over technical details, seemed to be converging toward a consensus. Arshad Kudrolli, a physicist at Clark University with an affable, chatty manner, had added two more equations to a slide. "That's what it amounts to," he concluded. "Then you can just, if you don't care about—"

"Yeah, yeah," Chris said. "I agree. I mean, I think that's great."

"So, small *n*—it's tom-a-toes versus tom-ah-toes as far as this conversation is concerned, but at least we can see that that is the same," Arshad said. "So the question is, how to interpret *b*..."

On an overcast October day about a year earlier, during a trip to the Boston area, I'd taken the train into Cambridge and made my way to the Jefferson Physical Laboratory at Harvard. Everything on campus—the neat red brick buildings, elegantly curved trees, brilliant gold leaves drifting to the manicured grass, shafts of sunlight breaking through the clouds—oozed Harvardness, the illusion of intellectual nirvana punctured only by the loud, aggressive whine of a mower.

Madelyn, who met me by the front steps of the physics building, was a congenial guide to the world of physics and had a knack for explaining esoteric concepts. At one point, when discussing granular flow—any situation where items moved in distinct pieces—she said, "Like if you stay in a hotel, and they've got the twisty things to dispense the cereal, and then the cereal is always getting caught? That's a granular flow problem." She'd recently published a tongue-in-cheek paper with a friend about how to classify all foods as a soup, salad, or sandwich using a method similar to the one for categorizing other phases of matter (thus definitively settling the question of whether a hot dog was a sandwich; the answer was yes).

Her office was an attic-like space with a steeply sloping ceiling, long wooden desks along the sides, and a giant whiteboard. When Madelyn used to share the office with several other graduate students, the room got unbearably warm from the heat coming off their computers running intensive calculations. She spent her days scribbling equations, drawing geometric shapes, writing software, running simulations, and "staring out the window," she had told me earlier.

Over the last couple of years, Madelyn had been working on

a computer simulation of sheet crumpling, represented as a mesh of thousands of connected nodes. The nodes were connected by springs; at each point in time, the software calculated how each node responded when a force was applied. To make sure she could trust the computer model, Madelyn had been running tests on the virtual sheet—stretching, bending, pinning a weight on it—and checking whether it responded the way she'd expect based on theoretical calculations or compared to experimental data on real sheets.

With simulations, Madelyn hoped to learn something about paper crumpling that couldn't be easily gleaned through experiments. For instance, she could tweak properties like the thickness or stiffness of the sheet and see what happened, an exercise that would be much more laborious in a lab. She could see how energy moved through the sheet and perhaps eventually predict where crumples would form. "The way that we are able to make predictions in physics is really, at the end of the day, by looking at energy flow," she said.

Madelyn had started out studying astrophysics and cosmology, investigating theoretical particles called axions that might make up dark matter. Axions could form large clusters in space, and she'd tried to figure out how big those clusters would get before they became unstable. Her result contradicted a well-established scientist's conclusion: He had claimed that if the cluster got too big, it would collapse and form a very small black hole, while Madelyn's calculations suggested that the cluster would explode. That was when she started to feel like a scientist. She'd proven someone else wrong; she could contribute something.

Madelyn eventually landed on soft matter physics—a field encompassing any material that "would have some kind of squishy reaction" if you poked it. "I was interested in how bubbles form on the surface of my coffee and how paper crumples and how you turn oil into mayo," she said. "These are the kinds of questions that haunted me." And a lot of hard problems remained to be solved in the field because these squishy behaviors "are all the things that most physicists hate," she said. They were messy and unpredictable, not neat systems like a pendulum swinging reliably back and forth.

She still saw links between her current work and past research in cosmology. The common theme, she explained, was small structures that gave rise to big phenomena. For instance, little fluctuations in the density of dark matter helped drive the formation of stars, heavier elements, life. And when you crumpled paper, the little folds made the sheet rigid; it no longer flopped over when you held it up. Minor changes in the physical system had translated up to create an entirely new material. "Paper crumpling is especially special to me because I can see all the ways that it connects from a nanoscale to a millimeter scale to a parsec scale," she said. "It's been really cool to see those threads of the physics that unite the whole universe."

Interest in crumpling went back hundreds of years, to artists who depicted draped fabric. Imagine Leonardo da Vinci painting the *Mona Lisa*, said Tom Witten, a professor emeritus of physics at the University of Chicago: "Just think how carefully he's trying to do the folds on her sleeve." Though the folds weren't as sharp as those produced by paper crumpling, they were the same kind of "singularities," he said. "He was trying to get the phenomenon right by reproducing it in drawing. And we try to get the phenomenon right by reproducing it in math."

On the science side, early physicists and mathematicians were investigating the mechanical properties of materials—that is, when they would bend, stretch, or break in response to forces. For instance, what would happen if a large weight were attached to the end of a beam? If a rod were propped between two walls that pressed together, when would it buckle? (Imagine the scene in *Star Wars* where Luke, Leia, Han Solo, and Chewbacca are trapped in the trash compactor room, trying to save themselves by bracing a long pole between the walls closing in.) Or how much force would it take to deform concrete, wood, or steel parts of a particular size or shape? The answers would become useful when people had to design buildings, bridges, and other structures.

But what about thin sheets? Here things got a bit hairy. Materials like concrete or wood were relatively easy to understand; if you pushed or squeezed them, they didn't move much. But a thin sheet was much more sensitive. "I mean, the paper's going every which way," said Anne Meeussen, a physicist at Harvard. If you tried to create a computer model of the process, as scientists had done for typical building materials, "things become very complicated very quickly." You had to divide the sheet into small quadrants and calculate how each one behaved and how they influenced one another: "You don't just have a million quadrants, but also then a million million interactions," she said.

These questions were part of physicists' larger interest in complex systems, which meant many simple components behaving together in a not-so-simple way. Take sand, for example: a collection of little particles, each of which is easy to understand. But when you put them together, strange things happen. You can pour sand like a liquid, but when you walk on the beach, it supports you like a solid; yet sand doesn't conform to the typical definitions of either a liquid or a solid.

In the 1990s, some scientists started delving into the theoretical aspects of sheet crumpling. Martine Ben Amar (the physicist in Paris) and a colleague described a feature called a d-cone, which emerged if you poked a sheet into a confined space and made some material buckle. This was arguably the simplest example of a point singularity, Witten said. Meanwhile, Witten's team investigated the properties of ridges and could predict, for instance, how sharp a ridge would be given the sheet's thickness.

Lab experiments on sheet crumpling yielded unexpected findings. In one study, Menon and a colleague crumpled aluminum foil into balls and ran them through a CT scan. They found that the density of material didn't change much between the balls' interiors and edges. "It's surprising, right?" Menon said. "You squashed it from the outside." Nor was there much difference in the number of highly kinked spots or the orientation of the sheet's layers. If you were a lilliputian creature inside, trying to escape the labyrinth of creases, you wouldn't be able to tell if you were going in the right direction based on these geometric features.

In another study, Nagel's team had crumpled a sheet of Mylar, a thin plastic material, into a ball and placed it in a cylinder under a piston with a weight on top. They wanted to see how much the ball compressed, and they thought they'd measure the change in height once and be done. Instead, the piston kept slowly smushing the wadded sheet, the height lowering in a logarithmic pattern without reaching a stable stopping point. The sinking just went on and on for weeks, a result that Nagel found "totally bizarre."

About a decade later, physicists Shmuel Rubinstein and Omer Gottesman undertook a similar set of experiments at the Weizmann Institute of Science in Rehovot, Israel, and at Harvard. Omer squished Mylar sheets with a piston in a cylinder, but this time he scanned the crumples afterward with a laser. Using image-analysis software, he identified all the creases and added up their lengths. Then he put the crumpled sheet back in the cylinder and repeated the process, over and over.

Again, a clear pattern emerged. The total crease length increased logarithmically, meaning that it initially rose quickly and then kept rising at a slower rate. At the time, Madelyn's adviser, Chris, worked across the hall, and he was fascinated when Shmuel sketched the findings on a blackboard in his office. "That's really quite a remarkable result," said Chris, who became a collaborator on the project. When you see very disordered systems, "it seems all random. You wouldn't expect that you could say anything—like *anything*—definite about this."

Another peculiar finding had emerged as well. The graph of total crease length followed the same type of curve regardless of the sheet's "history." For instance, imagine if you crushed a sheet tightly in your hands once and then gently crumpled a second sheet ten times, so that both sheets ended up with the same total crease length. If you then stuck both sheets into cylinders and crushed them with pistons, their crease length growth would follow the same logarithmic pattern, even though they'd initially been handled in different ways. "When I first learned about that, I was really mystified about why that should be true," Chris said.

It was striking that they didn't have to track details such as the locations of folds or how they affected other folds to figure out how much crease length would increase, said Ian Tobasco, an applied mathematician at Rutgers University in Piscataway, New Jersey, who wasn't involved in the research. "It's surprising to learn that you can ignore all that, and you get an accurate prediction of the total length," he said.

But the study cried out for an explanation. Anytime you see a logarithmic pattern like this, "you sort of sit up and say, 'Huh,'" said Andrea Liu, a physicist at the University of Pennsylvania in Philadelphia. "There's some process that really needs to be understood here."

"If you look at it node-based, then yeah, you're going to have intuition problems. But if you look at it edge-based, every triangle in the mesh is the same."

"I guess I feel that, looking at that four node, it's a case where you're not seeing enough of the elephant to make a decision about—it's like you've only got one leg of the elephant."

"True."

"I did wonder whether a triangle is enough…"

I was on a Zoom meeting about Madelyn's crumpling simulation, and the team was deep in a geometrical discussion of something called the Warren-girder case, which she later explained to me meant a mesh of squares divided into right triangles. On the call was Chris, working from a cluttered room in front of a couch covered with cushions, strewn clothing, and two white teddy bears placidly observing the conversation. He had an angular frame and a penchant for shirts and scarves in brilliant hues like mulberry and cherry red; today, he was sporting a canary-yellow T-shirt. In a photo I later saw of Chris's research group, about a dozen vibrantly dressed team members were arrayed in rainbow order on the steps of a building: Madelyn in pumpkin-orange, Chris in marigold (and a cheeky grin), and others in scarlet, lemon-yellow, jade green, cerulean, lavender, and plum.

Madelyn was running into problems with simulations of shearing—a motion similar to grabbing the sheet with both hands and pulling one hand upward and the other downward. "It scales roughly the same way we were expecting," she said. "But as you can see, the magnitude of the error is about a hundred to a thousand times larger. And I am not sure why."

"That's strange," Chris said. "Strange."

Jovana Andrejevic Kim, another PhD student on the team at the time, had also joined. "Could I ask really quickly, what is your stopping condition for each of the runs?" she said. "It just crossed my mind as, maybe, being a slightly different case—it might take a bit longer to actually equilibrate." As part of the piston-crumpling study, Jovana had confirmed Omer's initial results using a second analysis method. Originally from Serbia, she had a neat, studious appearance and a tendency to preface her comments with disclaimers such as "I'm not sure if this might be useful, but..." despite having, at least to my layman's eye, an intellect so off the charts that she almost seemed to belong to a more advanced evolutionary stage of *Homo sapiens*.

Several years ago, Jovana had started investigating the reason behind the logarithmic pattern. The question was, "Why is this happening?" she said. "We wanted to back it up by something that was physical." Shmuel had suggested that she look at the facets, the shapes delineated by creases. A computer algorithm wasn't reliable enough at identifying facets, so Jovana undertook the herculean task of outlining them manually, tracking how new ones emerged with each round of crumpling. At one point, she was spending at least five or six hours a day on hand-tracing; one sheet had nearly four thousand facets. "I didn't actually realize how much work it would amount to," she said. If she had, "I still would have done it."

Chris sometimes worked at a research institution in California, bringing students with him during the summers; while they were there, he and Jovana hiked on the weekends. Along the gravelly paths, they saw rocks that had broken into assorted sizes. Jovana had also started reading about fragmentation theory, equations that researchers used to explain the fracturing of materials into separate objects—shards of glass, for instance, or pieces of pumice and ash from volcanic eruptions. She found that the same theory could predict how the assortment of facet sizes on the crumpled sheets—that is, the number of facets of each size—would likely change over time. When new creases appeared, it was as if a large facet had broken into smaller ones.

That was "a brilliant intellectual leap," said Andrea Liu, who later hired Jovana as a postdoctoral researcher at the University of Pennsylvania. The connection is "not at all obvious." When you think about crumpling, she said, "you don't think about a solid breaking up into pieces."

In the team's paper, Jovana had created meticulous diagrams of a sheet as it became more and more crumply. In the first picture, the sheet was broken into large fragments delineated in shades of blue, purple, and burgundy. Over the next few pictures, as if you were turning a kaleidoscope, the big facets shattered into progressively smaller pieces, until you were left at the end with an elaborate mosaic of tiny shards.

The study helped explain the logarithmic growth, the team thought. When you crumpled a sheet repeatedly and the facets got smaller, the sheet also became more pliable. So fewer new creases formed, and the rate of crease formation slowed down.

In a later Zoom meeting, Chris explained the piston-crumpling and facet-size studies to Arshad and Amit Dawadi, a PhD student in Arshad's lab. "One thing I found really tantalizing," Chris said, "is the fact that all of those statistics that we have for the original piston crumpling—seems a lot of that, it's clear that it holds fairly broadly." His team had run a computer simulation of a sheet crumpling into a ball and tracked the facet sizes that emerged. "That's a totally different way you can crumple a sheet," he said. "But I think the thing that's really tantalizing"—he kept repeating that word—"is that if you look at all of those statistics, they're the same." And in a twisting simulation, the facet-size patterns in the center of the sheet were, again, similar to those in the piston data.

Chris proposed that the team repeatedly crumple sheets using both lab experiments and computer simulations. After several rounds of crumpling, "I suspect we'll see this logarithmic growth emerge in many different scenarios," he said. Later, he expressed hope that the team could come up with rules that captured the sheets' behavior—for instance, the relationship between the amount of compaction and how quickly crease length increased—in a universal way. "What would be really,

really awesome," he said, "is if there's a more general law here that connects…everything."

I arrived at Clark University on a sunny, blustery fall day. The campus occupied a few square blocks in the middle of Worcester, a previously industrial city west of Boston; Arshad's lab was in a slightly incongruous modern glass structure tucked between two brick buildings. On the second floor, I found what Arshad described as a "typical chaotic physics lab," a fluorescent-lit space where nearly every surface was crammed with equipment, materials, and tools. Among the various items were an ultrasonic cleaner, Febreze, Stretch-Tite plastic wrap, Karo light corn syrup, Crayola modeling clay, black wires hanging off the walls like vines, a long cardboard box with "LASER" written on it, and shiny blue tins of Royal Dansk butter cookies (containing, I guessed, something other than cookies).

Despite having been a professor for nearly three decades, Arshad still appeared to brim with youthful delight in science. "Our goal, in some sense, is to be like a child," he said. He wanted to be "really pure about it and just think in terms of knowledge." Equations didn't speak to him; during one of the team's discussions of a calculation, he claimed, unconvincingly, that "my math skills are zero." He had to do experiments with physical objects, to pick things up with his hands. "I'm not the person who's, you know, painting the Sistine Chapel," he mused, when describing what type of scientist he was. He wasn't quite even Matisse; he thought he was more like Marcel Duchamp, the French artist who famously created the work *Fountain*, a urinal with a signature. "He's the one who puts the commode up there and says, 'That's art.'"

Once Arshad started talking about his team's work, I could barely keep up. He said something about thinking of liquid like little marbles, going "all Dungeons & Dragons" and buying a lot of dice, and what would happen if you bent an uncooked spaghetti strand (a "quasi-one-dimensional object"); when Madelyn arrived at the lab for a visit, they started spitballing about quasicrystals, complex geometric patterns that had been found in ancient Islamic tile mosaics. But for Arshad, everything came back to a common thread: "How do you get from something which is ordered and all put together, to something which is fragmented and disordered?" he asked. When they crumpled a sheet of paper, did it "break" in a predictable way? Answering these questions was like climbing "a thousand-foot stairwell" in *The Lord of the Rings*, he said. "I feel like I'm always halfway."

Trudging up the stairwell was Amit, the graduate student doing the experimental labor. I'd met him a couple days earlier to see his paper crumpling endeavors; he'd immediately gotten to work, slicing a sheet of paper with a paper cutter into rectangular segments. His experimental setup was in a corner of the lab, enclosed by medical-looking green curtains. On a metal surface about the size of a pool table, perforated with a grid of holes, was a rod between two clamps, with white plastic discs attached. Amit rolled the paper into a cylinder between the two discs, taped the seam, and tightened thin metal circles, like bracelets, over the ends of the cylinder.

At a computer, he controlled how much the sheet should be twisted, while a camera above the table recorded images of the crumpling. The right disc started rotating jerkily, driven by a motor; the left clamp slid along a track as the paper became more and more twisted, creating a flower-like pattern of creases. Amit repeated the experiment with more sheets, peering at the pictures of crumpling on the computer like a doctor trying to diagnose a mysterious illness on an X-ray. Even though the work seemed tedious, he got more and more animated as the morning progressed, scurrying between his computer and a yellow legal pad and the paper cylinder, tweaking the software, scribbling numbers, adjusting the experimental setup, and murmuring numbers under his breath.

After crumpling several sheets, Amit was ready to scan them with the laser. We moved to a cramped room with a computer, lab bench, sink, and camera on a tripod with a large sign entreating, "Please, Don't Touch the Camera Stand!" On a shelf, he had stored an Amazon box filled with crumpled papers of different colors and thicknesses. At one point, I investigated a mysterious dish filled with small pink blobs on the counter; as I moved closer to take a picture, I heard him say politely, "There is a camera stand." I didn't register why he was mentioning the camera and said "Okay," obliviously. Then I realized that I was precariously close to one tripod leg, and what Amit actually meant was "STEP AWAY FROM MY CAMERA *NOW*." "Sorry sorry sorry sorry!" I said, backing away.

Amit taped a crumpled sheet onto a black platform, with a thin seam of red light shining on it from the laser overhead. He turned off the room light, and the platform started moving slowly, the red line inching across the surface. The camera would capture 620 frames of the sheet, about a tenth of a millimeter per frame. "I didn't want to miss any information," he said.

As the laser scanned the creases, Amit showed me previous scans on the computer: intricate landscapes of blue, green, and gold. “Ridge, valley, ridge, valley, ridge, valley,” he said. “That gives me the surface.” They could determine the curvature, slope of the ridges and valleys, and three-dimensional coordinates of every point. I said it looked like a mountain range, and he got excited. “Exactly the same! Exactly!” Researchers had used a similar technique called LiDAR (Light Detection and Ranging) to map the topography of forests: They typically flew overhead in a plane or helicopter, shone a laser downward, and captured the light bouncing back to measure factors such as elevation or canopy height.

So far, Amit had crumpled more than one thousand pieces of paper with different thicknesses, sizes, and degrees of rotation. When I asked him if he ever got bored, he said no, because “you will get at least something”—some tidbit of data—“when observing things.” I wondered aloud what people thought when he told them that he crumpled paper for a living. “Sounds funny, but it’s kind of interesting, yeah?” he said. Taking a sheet of paper, “twisting it here, there, all the time, right?”

A few months later, the team convened over Zoom to share their latest updates. Arshad said that he and Amit had been trying to answer Chris’s question: When they folded a sheet of paper “once, twice, four times, what creases do we see?”

Well, they weren’t seeing the expected logarithmic growth. In a graph that Amit shared, the total crease length stayed about the same, regardless of whether a sheet had been twisted once, twice, or even ten or eleven times. "We can't see any evolution," Arshad said. It seemed like the sheet kept refolding along the same lines, like origami. He suspected that their automated analysis method wasn't detecting all the creases; after a certain amount of twisting, he said, "it's clearly missing some of the subtle features." He and Amit would work on improving the software's ridge detection. "I'm not saying I totally believe this yet," Arshad said.

Things started turning around in the next meeting. In previous experiments, Amit had untwisted the sheet between each crumpling cycle, but he hadn't taken it off the apparatus. Now, he was removing and unfurling the sheet, and sometimes flipping or turning it, before retwisting. These extra steps might allow creases to flatten or pop in the other direction, making it harder for the sheet to repeat the same folds. He'd also tried using Mylar, which was tougher than paper.

When Amit showed his new crease length data, the curves seemed promising; they looked closer to the pattern seen in the piston experiments. "I mean, that looks a very convincing logarithm," Chris said.

Arshad laughed. "Yeah, I mean now you are getting serious," he said. If the graph had a few more data points, "even the skeptical experimentalist in me will start believing you." He thought Amit should perform five additional crumpling cycles. "I'm not totally on board yet," he said.

"But it is very interesting," Chris persisted.

"It's nice to see. I mean, don't get me wrong," Arshad said. "I think we had, till three days ago, no idea that this would happen." Toward the end of the meeting, he acknowledged, "My faith in all of what we're doing has increased tremendously." With Amit's new data, "it gives confidence that we are at least in a position to see it."

"I'm very optimistic," Chris said. "I just feel we've really gained some real understanding there of what's going on in these systems—that there's some universal underpinning to how damage accumulates." If Madelyn saw a logarithmic increase of similar magnitude in her simulations, Arshad concluded, "then we have a solid connection."

But Madelyn had reservations. When I asked her a few months later if Arshad had become more convinced of the log growth, she said, "I think he's really been won over." Meanwhile, she'd flipped in the opposite direction. "I've become a lot more skeptical," she said. "It isn't universal."

In an essay published in 1907, philosopher and mathematician Bertrand Russell famously wrote, "Mathematics, rightly viewed, possesses not only truth, but supreme beauty—a beauty cold and austere, like that of sculpture, without appeal to any part of our weaker nature, without the gorgeous trappings of painting or music, yet sublimely pure, and capable of a stern perfection such as only the greatest art can show." When I looked at the team's piston-crumpling paper, the orderly graph of crease length growth seemed to capture the qualities that Russell had described. The points fell tidily along one curve, almost as if drawn to it by gravity; randomness was transformed into predictability, the messy and turbulent distilled into something elegant and pristine. And Amit's results had followed a similar pattern—evidence of universality, that prize that physicists strive for, the laws that transcend details. When those dots lined up in just the right way, that meant, Yes, you've understood; you've struck the underlying truth of the matter.

Now, Madelyn had found something that was perhaps true, but not beautiful. Her computer simulations weren't fully conforming to the pattern that Omer and Amit had found in their lab experiments. When she twisted a virtual sheet like a candy wrapper, Madelyn did see the log pattern. But when she simulated crushing a sheet into a ball, the results were messier. The expected log growth emerged only if the sheet was heavily crumpled—that is, when the amount of "compaction" passed a threshold.

If you had asked her a few years ago what she expected to see, "I would have said, at all compactions, we're hoping to see the exact same law of growth," she said. "That's not what we found." In other words, the universal law came with caveats; it emerged only in certain scenarios.

A second, more nagging problem had come up too. Even when crease length did grow logarithmically in the ball-crumpling simulation, it eventually plateaued instead of continuing to increase. Madelyn had tried running the simulations in higher resolution—that is, using a virtual sheet with more nodes—but that hadn't fixed the problem. When she showed the group her latest graphs, Arshad said, "To be honest with you, Madelyn, I'm just going to be the annoying guy. I mean, to me, it looks like, okay, there's some transient—and then it saturates." In other words, the crease length rose for a bit and then flattened out.

"I guess I'm just a little surprised that it would do that," Chris said. His brow was furrowed, and he'd sunk deep into his magenta puffy coat, the collar up around his ears.

The team had been planning to report their new results on twisted sheets in a scientific paper, but Arshad expressed doubt that the ball-crumpling data should be included. "Right now, we are in a situation where it doesn't progress the same as the experiments," he said. "Every experiment is showing this log growth." And they didn't fully understand what was going on. It was possible that the plateau wasn't a real result, per se, but an artifact due to some technical issue—a need for even higher resolution, perhaps, or the software incorrectly deciding what counted as a crease.

Later, I asked Madelyn if she worried that the team was ignoring something important by leaving out the ball-crumpling data. "It's hard to separate the scientist from the science," she said. "I am a scientist with feelings and with intuitions." She thought that her finding didn't necessarily contradict Amit's data. In her mind, the simulation actually provided more evidence of what they already knew—that a certain amount of "noise" or disorder had to be present to get the logarithmic pattern. For instance, in the twisting and piston experiments, the sheets had been flattened, flipped, or turned between crumpling cycles. But in her ball-crumpling simulation, the sheet hadn't been manipulated after each crumpling round.

In other words, the plateau might mean that the team needed to add a second, unbeautiful caveat to the law of log growth. "Maybe Chris's initial gut intuition was that 'This is universal, full stop,'" she said. "I think we've now modified it to be, 'This is universal, comma, when you have a high enough compaction and enough noise.'" So not sublime purity, as Russell had put it. Over the next month, Madelyn's further tests of the simulation suggested that the plateau wasn't just a technical artifact; it seemed to be a real result.

I read the Bertrand Russell quote to her and asked what she thought—about the idea that adding caveats to the log growth rule made it more true, perhaps, but less pretty. She said, "This is why I love physics." She'd often heard her professors say about other theories, "It would be nice if this model worked out, but that's just not the universe we live in." Perhaps you started with a simple equation, but you couldn't make it fit something as messy as real life, and she liked that element of uncertainty.

She thought it was funny that she felt that way, because she was otherwise a perfectionist: If she could guarantee that her bookshelf would always look tidy, that she would always be on time, that she would never say the wrong thing at dinner parties, and that she could generally control everything, she would. "Maybe I find it comforting that we can't understand everything," Madelyn said. In fact, the world as we knew it had only come into being because of quirks and inconsistencies. Slight fluctuations in the density of dark matter were likely crucial to the emergence of stars and planets, including Earth. In a hypothetical alternate world where everything always stayed perfectly uniform, none of that would have happened. "I guess I'm not looking for austerity," she said. "I'm looking for a little bit of chaos."

The stern perfection that Russell described made her think of artwork by seventeenth-century Dutch Golden Age painters, some of whom depicted scenes in incredibly precise and exacting detail. Maybe the reality of our messy universe was more akin to the Impressionists' paintings—landscapes of lilies and sunlit waters rendered in soft dabs of color.

"They can both be beautiful," she said. "They can both make you feel something."

COVID-19

Molecular Biology

Bright Lines

"Is anyone else starting to feel more than a little apprehensive about this virus?"

England

In December 1832, a shoemaker's wife in Leeds, England, died of cholera. Her family mailed her clothing to the woman's brother, a farm worker named John Barnes, presumably thinking it might be of use. The clothing was unwashed, and shortly after opening the box, Barnes fell ill and died. His mother-in-law, who cared for Barnes's wife when she also became sick, collapsed while walking home; she then infected her husband and daughter, and the entire family succumbed to the disease.

These cases were recounted in the second edition of the book *On the Mode of Communication of Cholera*, published in 1855 by the now-legendary epidemiologist John Snow. The report was full of death, near misses, hubris, tragedies upon tragedies. In another town, residents of houses complained to the property owner about their contaminated well water. The owner smelled the water and said "he could perceive nothing the matter with it." Asked to taste it, he did, and he died of cholera three days later.

Snow's most famous contribution was his investigation of the outbreak around Broad Street in London, which he deduced was caused by people drinking from a pump. He relentlessly chronicled the deaths of patients. There were nine customers of a coffee shop who were served pump water and died. There was the twenty-eight-year-old pregnant woman who fetched water from the pump and died two weeks later. There was the man who traveled from Brighton to visit his cholera-stricken brother, who died before he arrived; the visiting brother stayed a mere twenty minutes to eat a "hasty and scanty luncheon of rumpsteak" and a "small tumbler of brandy and water" and died two nights later. There was the percussion cap–maker's wife who had been given a bottle of water from the pump and died, and her visiting niece who drank the water and also died. There was the tailor, the army officer, the cabinetmaker. It was, Snow wrote, "the most terrible outbreak of cholera which ever occurred in this kingdom."

Snow kept meticulous tables of data, an endless catalog of cases, and a map marking deaths in the neighborhood surrounding the pump, a small black bar at each address where someone had become fatally infected or perished. In part of his report, he documented his painstaking efforts to determine if contaminated water supplied by a certain company was to blame for another cholera outbreak. "I resolved to spare no exertion which might be necessary to ascertain the exact effect," he wrote, for the issue was "of so vast importance to the community, that it could not be too rigidly examined, or established on too firm a basis."

Washington

It was May 2020, two months into lockdown. I was on Zoom (of course) with Louise Moncla, who was then a postdoctoral researcher at the Fred Hutchinson Cancer Center in Seattle. She was quarantining in her apartment on Capitol Hill, a trendy neighborhood where the normally bustling bars and restaurants had emptied.

Today, Louise was analyzing a new set of genome sequences of SARS-CoV-2, the virus that caused COVID-19. For the last several months, scientists around the world had been extracting the virus's genetic material from patient samples: the long chains of molecular instructions that dictated SARS-CoV-2's structure and how it infected humans and other animals. Because the genome was mutating over time, the sequences held clues to how outbreaks had unfolded. Based on minute differences between each patient sample, researchers could infer the chain of transmission.

That day, new sequences had been posted from Osmania University in Hyderabad, India, and Karolinska Institutet in Stockholm, Sweden. Yesterday had been quiet until about 5 p.m., when about two thousand sequences poured in, mostly from England. Louise clicked on one earlier entry: a sample collected by nasal swab on April 6 in Montreal, Quebec, submitted by a lab at the university hospital center CHU Sainte-Justine. The patient information was listed as "Gender: unknown. Patient age: unknown. Patient status: unknown." I wondered what had happened to them. One line summarized, succinctly, the problem with the pandemic. It read: "Host: Human."

When Louise opened a sequence from a patient in Algeria, I squinted at the screen. The virus's genetic code was made of four molecules called nucleotides, each designated A, T, C, or G. The genome began "CTTGTAGATCTGTTCT…" and continued for about thirty thousand more letters—around one hundred thousand times smaller than the roughly three-billion-nucleotide human genome. Though I'd worked in molecular biology labs for years, I still found it hard to understand how such a tiny piece of genetic code could wreak such havoc. Economies had crashed, social ties had crumbled; by the end of that month, the death toll from COVID-19 in the United States would reach one hundred thousand. "That's actually one of the reasons I decided to study viruses," Louise said. They're so small, "but they can just completely shape everything about our world." Back in February, she had emailed her friends to prepare them for what was coming. She said that she thought a pandemic was going to happen, and it might really change things.

It had started with a tweet on January 11, 2020, at 1:08 a.m. GMT, which one scientist told me would "go down in history as probably the most important tweet." Edward Holmes, a virologist at the University of Sydney in Australia, posted a link to a publicly available genome sequence of the SARS-CoV-2 virus, produced by virologist Zhang Yongzhen's team at Fudan University in Shanghai, China. The sample had been collected from a forty-one-year-old worker at a Wuhan seafood market, who was admitted to the hospital with fever, cough, pain, and a tight feeling in his chest. Over the next sixteen months, scientists went on to sequence more than a million virus samples, often in near real time—something that had never happened before so quickly.

As those sequences poured out of labs, Louise's team was just as rapidly analyzing the data and displaying the results on a website called Nextstrain. Every day, the researchers ran software that arranged virus samples into a tree, with the root at the left and branches stretching horizontally to the right. Each branch represented a new version of the virus with a different set of mutations; colored dots on the lines and tips represented samples from infected people. The trees captured not just the major variants that became familiar over the pandemic—Delta, Omicron, and so on—but many other minor variations circulating around the world.

In an early report posted on January 25, 2020, the first U.S. case on the tree was marked with a red dot. It was from my home state of Washington, a traveler who'd recently returned from Wuhan. I have no recollection of any feeling of alarm from that time; COVID-19 still felt far away. My first indication of real concern was an email written on February 25 from another writer to a group of colleagues: "Is anyone else starting to feel more than a little apprehensive about this virus?"

Four days later, news outlets reported that the first known COVID-19 death in the United States had happened at a hospital about five miles from my home in Kirkland. When it became clear shortly afterward that an outbreak had erupted at a local nursing facility, the *New York Times* described Kirkland as an "epicenter of both illness and fear," not something I'd ever expected to read about our low-key suburb.

As 2020 dragged on, and as reality set in and refused to budge,

I heard about the number of SARS-CoV-2 sequences in a global database climbing from a few dozen, to hundreds, to tens and then hundreds of thousands. Working from home, feeling like I ought to do something, I spoke to researchers who were scrutinizing this data to better understand the virus's spread through their state, their country, or around the world. Some were from Seattle, just across Lake Washington from me; others worked in Wisconsin, the West Bank, Uruguay, the Democratic Republic of the Congo, New Zealand. They told me about cases spreading through a daycare, a college town, a nursing home, a quarantine hotel.

When I looked back at the first Nextstrain reports about the pandemic, the descriptions already felt remote, like passages of a history textbook about a catastrophe that had befallen an ancient civilization. "In December 2019, a new illness was first detected in Wuhan, China," the team wrote on January 23, 2020. "We now know this to be another outbreak of coronavirus in humans…"

US and Switzerland

Miguel Paredes's medical school class had been studying the gut when the seriousness of COVID-19 became clear to scientists in early 2020. Miguel was an MD-PhD student at the University of Washington in Seattle, whose family had moved from Colombia to the United States under political asylum when he was about five years old. He originally thought he would study neuroscience in college, but he abandoned it because the field felt more like a philosophical pursuit, almost selfish. "It was really these beautiful questions about who we are and how do we come to be," he told me. "But there's people that need help now." After switching his focus to public health, he'd conducted research on the Zika pandemic in his hometown of Cali, Colombia, and decided to become a doctor.

Now he felt useless. He contacted Trevor Bedford—a computational biologist at the Fred Hutchinson Cancer Center, whose lab Miguel had worked in the previous summer—to ask if he could help. Bedford's team and their collaborators at the University of Basel in Switzerland had previously developed Nextstrain to analyze genomes of other viral pathogens and were now applying it to SARS-CoV-2. After moving to his parents' house in Florida, Miguel settled into a cycle. He'd wake up around 8 a.m., run Nextstrain on data that had come in overnight, and do his med school classwork in the afternoons while Louise took over on the West Coast. In the evenings, a collaborator in Switzerland stepped in, and Miguel stayed online to help until around 1 a.m. He spent most of the day and night in his room, coming out only for food.

The way that Nextstrain and similar genetic analysis software worked, as one team member described it, was like a game of kindergarteners playing telephone. Imagine a group of children sitting in a circle. You whisper "THAT" in the first child's ear. They whisper the word to the next kid, who whispers what they hear to the next kid, and so on. Along the way, the word subtly morphs—perhaps to "HAT," then "AT," then "CAT," and then "CAP."

You ask each of the children to write down the word they thought they heard on a scrap of paper. They drop the papers into a hat, and you shake them up and dump them out. Based on the similarities between words, you can make some reasonable guesses about which child was sitting next to which, even if you weren't in the room. For instance, the kid who wrote "CAP" was probably closer to the kid who wrote "CAT" than the one who wrote "HAT."

The same logic applied when examining changes to the SARS-CoV-2 genome. Every couple of weeks, the sequence mutated at one spot. For instance, the segment ACCG might change to ACAG. Next, the segment CTCA might change to CTCG, and so on. Nextstrain could reconstruct which versions of the virus had "descended" from which, and thus the order in which infections had spread among patients.

The picture was still incomplete: If several people were infected with exactly the same version of the virus, scientists wouldn't be able to tell who transmitted it to whom. And some links in the transmission chains remained missing. But combined with other methods such as interviewing people about their activities and close contacts, researchers could piece together what had happened.

The practice of reconstructing disease transmission has a long history. In a public health report published in England in 1905, I found what looked like an early precursor of a Nextstrain tree: a diagram showing how Nellie I., a five-year-old girl in Woolwich, UK, spread the measles to other children in her social circle. In the picture, a series of boxes—each containing the name of a child and their age—were connected by lines indicating who had likely infected whom. From Nellie, lines were drawn to her cousin Charles, her "bosom friend" Gertie, her "other bosom friend" Lily, a younger child in Lily's care named Sammy, another friend Ellen, and Nellie's newborn brother, Henry George. Meanwhile, Gertie gave the disease to her brother Dick; Charles infected his older sister Annie and next-door neighbor Chris; and Annie, in turn, passed it to her younger siblings Daniel and Andrew. The author of the case study, Dr. Thomas, wrote, "This popular young lady, Nellie

I., was therefore directly responsible for no less than 11 cases, which could be definitely traced to her."

But the Nextstrain diagrams soon grew far beyond anything Dr. Thomas could have envisioned. The first trees were neat and contained, with samples from only a couple dozen patients. Instead of names like "Nellie" and "Gertie," each dot was labeled with codes such as "Wuhan/IVDC-HB-01/2019" and "Guangdong/20SF040/2020." Then one version of SARS-CoV-2 with the same mutation showed up in Singapore, Hong Kong, France, and California; it was transported to Italy, mutated again, and hopped to England, Brazil, and Switzerland; another version infected people in Australia and Canada. The number of sequences climbed, first slowly, and then quickly. By late March, the tree was an impenetrable thicket.

Wisconsin

During the early days of spring 2020, the streets in Madison, Wisconsin, in the morning were foggy and empty. Gage Moreno drove his baby blue 2010 Mercury Milan to work—not the color he would have chosen, but his grandparents had given it to him because they didn't want him to take the bus. His partner and all his friends were working from home. Every weekday, Gage and fellow graduate student Katarina Braun headed to a basement lab at the University of Wisconsin–Madison to sequence SARS-CoV-2 samples from the hospital.

Gage and Katarina's work had started on Valentine's Day, when they analyzed the virus from the first COVID-19 case reported in Wisconsin. This was before they knew how bad it was going to get; they were excited to be among the first scientists in the US generating genome data. They didn't realize that they'd be sequencing samples all day, every weekday, for the next fifteen months.

At least once a week, they drove to the hospital. Someone from the diagnostic lab came out with a biohazard bag of tubes containing nasal swab or spit samples from positive patients, encased in a Styrofoam container inside a cardboard box. When cases first started ramping up, each delivery held about a few dozen tubes. Then the bags started getting bigger and bigger, hundreds of samples a week.

Gage and Katarina transported the samples in the trunk of the car and carried them to a basement lab space, where they suited up in Tyvek jackets, masks, safety glasses, booties, two pairs of gloves. For each tube, they killed the virus with a detergent and isolated RNA, the genetic material. Then they set up chemical reactions to convert the RNA to DNA—the format that could be read by the sequencing machine—and crank out copies of the DNA. It sounded simple, just moving liquid into tubes, but Gage and Katarina had to focus; they couldn't let their minds wander. It was easy to mess up. They could accidentally put RNA into the wrong tube, use the wrong enzyme, set the wrong temperature, drop a sample. When they made a mistake, they knew it immediately. That was an awful feeling.

Once they'd prepared the DNA, they placed the samples onto a chip on the sequencing machine, an unassuming looking boxy device smaller than a microwave. The chip had hundreds of pores on the surface, with an electrical current running through them. An enzyme ratcheted strands of DNA through a pore; as each nucleotide passed through the hole, the amount of current changed. By detecting these small electrical fluctuations, the machine could "read" the sequence.

As the months wore on, COVID-19 exhausted the public healthcare system. Gage and Katarina had their coping mechanisms. If one of them was in a bad mood, the other person would snap them out of it. On a bad day, they'd eat Sour Patch Kids, M&M's, Swedish Fish, and Flavor Blasted Goldfish; on an especially bad day, Jimmy John's and Diet Coke with pellet ice. If they were feeling angsty, they'd play Beyoncé's *Lemonade*. If they had a lot of work to do, Taylor Swift's *Folklore*, loud, on repeat.

A couple hours to the west, in the fall of 2020, Paraic Kenny could see parties raging at the homes of college students. When he drove to his lab on the University of Wisconsin–La Crosse campus late at night, he passed houses with flashing lights and throbbing music; once, on the way home from work, he counted more than a dozen parties within three blocks.

Paraic, director of the Gundersen Medical Foundation's Kabara Cancer Research Institute, had started sequencing SARS-CoV-2 samples with a colleague on the side when the pandemic hit. The researchers wondered what would happen as colleges in their city reopened and thousands of students returned from their homes. To Paraic's surprise, the number of times the virus was carried to La Crosse by students appeared to be quite small. But the students' behavior, he said, then led to "massive amplification."

Based on genomic similarities from patient samples, the team concluded that most of the spread was driven by two versions, or substrains, of the virus. Those same versions later showed up among infected residents in long-term care facilities. "Then many of the patients in the nursing homes started to die," he said in

November 2020. At that time, thirteen deaths in those facilities were connected to the two substrains from college students. "The ability to use genetics to draw these bright lines between these outbreaks, I think, is really important," Paraic said. "It makes it much harder for people who would prefer to shrug their shoulders and obfuscate to walk away."

In the meantime, his team was continuing to sequence samples from people who were more exposed to the virus, such as health-care workers. Near the end of our call, I asked if there were any times when he'd be unavailable for follow-up questions. "No," he said. "Unless I'm in the ICU myself."

The West Bank and Uruguay

In other parts of the world, the hunger for information in the early days of the pandemic was palpable. "Nobody at that point knew what was the situation in Palestine," said Nouar Qutob, a geneticist at Arab American University in Ramallah, in the West Bank. "We wanted to show real data." Maria Victoria Elizondo, a biotechnologist then at the Asociación Española Primera en Salud in Montevideo, Uruguay, told me that her team wanted to know "about our population, our Uruguayan society," and there was almost nothing written "about this country in the literature or anywhere else."

Nouar's team wanted to investigate rumors that had been floating among the public. Because the pandemic hadn't been as severe in the West Bank as in other places, such as Italy and the UK, Palestinians speculated they had gotten a "better," less deadly strain of SARS-CoV-2. If a patient tested positive after getting infected in another country, they would often say things like "we must have gotten the bad one," Nouar said. Because of these rumors, she said, some people worried that vaccines developed elsewhere in the world targeted different strains and would be less effective against the virus when given to Palestinians.

Nouar and her colleagues had already purchased a new sequencing machine before the pandemic, but their university was in the West Bank, and the machine was near Tel Aviv. Someone needed to visit the company to be trained to use the sequencer, and the team's lab technicians couldn't go because Palestinians needed a permit to enter Israel, which was difficult to get at the time. So Nouar, who was from Jerusalem and had an Israeli ID, traveled to receive the training; the machine was finally shipped to her lab in July 2020, and the team confirmed that the virus samples from Palestinian patients did in fact have genomes similar to those in other countries. The genetic history suggested that SARS-CoV-2 had entered the country in early February, about a month before the first cases were reported at a hotel near Bethlehem, likely spread by Greek tourists. That was a lesson, Nouar said; policy-makers should not have waited for the first case to appear before ramping up testing for the virus. "It was unnoticed," she said.

Victoria, the scientist in Uruguay, normally analyzed cancer patients' genomes. When the pandemic started, her team started diagnosing COVID-19 patients. The virus samples were just sitting in their lab, and they thought, "We should do something with this," she said.

Victoria got in touch with New York University Langone Health's Genome Technology Center in New York City, where researchers were already sequencing the virus, and she shipped the RNA samples to them on dry ice. The New York team's lab was right by the hospital; they could hear sirens wailing all day and see the refrigerator trucks out the window of one of the science buildings. Dacia Dimartino, a molecular biologist at the center, told me that it had felt scary at the beginning, but she was glad to do the research, "because otherwise I wasn't able to do anything else directly to help the community."

After sequencing forty-four samples, the team concluded that the virus had entered Uruguay at least twenty-three separate times, mostly through people arriving from other parts of South America. One set of cases came from a hospital in the small city of Rivera on the border with Brazil; transmission was likely sparked by a metallurgic worker who commuted back and forth between the two countries. While the researchers were analyzing data, news reports were blowing up about one woman who had recently returned from a trip to Spain and attended a wedding in Uruguay with about five hundred guests. "Everybody was upset and mad," Victoria said. But their results showed that many other people had carried the virus into the country: "It wasn't just her."

The disgraced wedding guest reminded me of a case in sixteenth-century Italy, in a lakeside town then called Desenzano. That time, the scientist was a country doctor named Andrea Gratiolo di Salò. As described in the book *Cultures of Plague: Medical Thinking at the End of the Renaissance* by historian Samuel Cohn, Gratiolo was investigating a claim that one woman from a nearby town was responsible for bringing plague to the community. Gratiolo argued that the woman had traveled and slept beside eighteen other people in a small boat—one person even resting her head in the woman's lap all night—and none of her fellow passengers had come down with plague. Nor had the lap-sleeper's husband or four children, who shared a bed with her. "It cannot be said, that she had been the origin of the disease," Gratiolo concluded, and "opinions were so widespread that no one dared to seek out the truth."

Democratic Republic of the Congo

Placide Mbala-Kingebeni was accustomed to epidemics. When COVID-19 appeared, his country, the Democratic Republic of the Congo, had already been working to stem the spread of Ebola for decades. During previous outbreaks, Placide, head of the epidemiology and global health department at the Institut National de Recherche Biomédicale, and his colleagues had set up genome-sequencing equipment at labs in Kinshasa and Katwa. They had to contend with unrest at the latter site. Healthcare workers from the international organization Médecins Sans Frontières (MSF) worked at an Ebola treatment center next to the lab, and signs of hostility toward them had been rising. The MSF team was seen as connected to the government, which had blocked people in parts of the region from voting in the last presidential election; locals had also heard rumors that the virus was a fiction concocted by the West and that foreigners were conducting experiments on patients' bodies and extracting organs.

One night, a crowd of people threw rocks at the treatment center, set parts of the facility on fire, and tried to ignite its chlorine supply. After that, the scientists could enter the sequencing lab only if accompanied by armed escorts, for two-hour blocks. These restrictions, the team wrote in a paper, "provided insufficient time to complete steps of sequencing protocols."

Nevertheless, they'd gotten results. The researchers found, for instance, that motorcycle taxi drivers were likely spreading Ebola to passengers. Twenty patients who rode with one infected driver had the exact same virus sequence as he did. After that study, taxi drivers were prioritized for vaccination.

When I called Placide at his home in a quiet Kinshasa suburb, he had retreated to his porch to get some peace away from his four kids, the sounds of crying, and the TV. He explained that because of the country's experience with Ebola, they already had systems in place to identify infected people, trace contacts, and quarantine patients. Health workers visited communities to provide "le savoir," as Placide called it—information about the SARS-CoV-2 virus and what to do if they experienced symptoms. "We go to find the cases," he said. "We were not waiting for the cases to come to us." I asked what he thought of the United States' response to the pandemic, and there was a diplomatic pause. Finally, he said, kindly, "You are not used to facing this kind of outbreak."

His team had started sequencing SARS-CoV-2 samples in March 2020. The researchers wanted to show that the virus was real; many people in his country didn't believe that COVID-19 existed. Even politicians asked them, skeptically, "Is it true?" Eddy Kinganda-Lusamaki, a virologist at the institute, said. "And we were telling them, yeah, it's true."

When I spoke to another institute scientist, Amuri Aziza, in early 2021, she said their sequencing efforts had been hindered by a lack of reagents, the chemical substances needed to process the samples. This was an ongoing problem; the team didn't have a way to obtain reagents locally, and they usually had to rely on scientists entering the country to bring them. Cases were increasing, and she wanted to know if the current wave was driven mostly by local transmission or travelers from other countries. "We are still waiting for them," she said.

I asked how she had been personally affected by the pandemic, and she told me that she'd caught COVID-19 while responding to an Ebola outbreak in the northwest province of Équateur. Her case was mild, so it was "no problem," she said. At the end of our call, I asked for her official title at her institute. She said, "Just a biologist."

New Zealand

At first, the culprit was the trash can.

An infected person had somehow transmitted the virus to another guest staying next door at the Crowne Plaza Christchurch, a hotel converted to a managed isolation and quarantine (MIQ) facility. Nearly all travelers entering New Zealand by air were required to hunker down for two weeks at one of these facilities and test negative before going out in public. The two guests had never interacted, but they had both used a garbage bin in the hallway, and perhaps the infected person had left stray virus on the surface. "We think that the lid was the only point of contact that they had," Jemma Geoghegan, an evolutionary virologist at the University of Otago in Dunedin, said in December 2020.

Jemma's team had been assiduously reconstructing transmission chains from the start of the pandemic, hoping to guide public health measures. New Zealand had committed to a strategy of elimination: zero COVID-19 cases. The stakes were high if any cases seeped through the border. As one scientist put it, "One leaping of the fence to the community—that would be enough."

The case of the quarantine hotel residents had started in

September 2020, when a positive COVID-19 case appeared in Auckland. That patient, denoted only as "G" in reports, had recently flown from Delhi, India, spent two weeks in the Crowne Plaza MIQ facility, and then taken an eighty-five-minute flight to Auckland. When their close contacts on the flight were tested, two more positive cases turned up: D and E (a parent and their baby), who had sat behind G on the plane. D and baby E had stayed at the same quarantine hotel, and all three of them had tested negative before being discharged. The virus genome sequences from D and G were almost identical—but G's had one new mutation.

Meanwhile, another cluster of people—A, B, and C—had tested positive at the Crowne Plaza. One of them, patient C, had exactly the same virus sequence as D did. And C and D had stayed next door to each other in the quarantine hotel. Jemma's team therefore surmised that the transmission chain had gone as follows: C had infected D (the parent of the baby) in the hotel, who then infected G (the case identified in Auckland) on the domestic flight.

But how had C passed the virus to D? They could open their doors only for approved purposes such as bringing in delivered meals or going out to exercise at certain times. According to key card data and CCTV footage, they'd never been in the hallway at the same time. They'd both touched the lid of the hallway garbage bin, though—hence the public health team's initial line of thought that the bin was the source of transmission.

At the time, the standard explanation advanced by the World Health Organization (WHO) was that SARS-CoV-2 spread primarily through close contact and droplets. Someone who sprayed out large particles—for instance, while coughing or sneezing—could directly infect a nearby person. Another likely route, the agency said, was that the droplets could fall onto a surface that later transmitted the virus to people who touched it. Most people "were working with the spoken truth that SARS-CoV-2 was droplet-spread," said Sarah Berger, then nursing director at the Canterbury District Health Board's infection prevention and control service in Christchurch. Despite urging from many scientists, the WHO hadn't yet fully accepted the idea that the virus was airborne—that is, often transmitted through smaller exhaled particles called aerosols that could hang in the air for hours.

But Berger's team wasn't happy with the trash can explanation. Nearly an entire day had passed between the times the two guests had touched the bin lid. "I was just going, this can't be the way this transmitted," said Joshua Freeman, then the district health board's acting clinical director of infection prevention and control. An investigation into a separate cluster of cases among mariners, who had arrived from Russia and spread the virus to two nurses at another quarantine hotel, turned up more evidence that SARS-CoV-2 could be transmitted through aerosols. When an infected person opened their hotel room door, contaminated air flowed into the hallway. And corridors were often unventilated.

The team went back to the CCTV records and saw that a nurse had tested C and D on the same day, one after the other. "Bingo," Freeman said. C had been tested first; fifty seconds later, D opened their door with baby E on their hip. This, the team believed, was the point of infection: D pulled down their mask to be swabbed and breathed in the stale hallway air, now infused with virus-laden particles from the room next door.

In some ways, the WHO's slow acceptance of airborne transmission echoed the resistance that John Snow met during his cholera investigations in nineteenth-century England—but it was a story played out in reverse. During Snow's time, the prevailing view among doctors and London's public health officials was that cholera spread through the air. People became infected when they breathed in a foul miasma, the thinking went. Despite Snow's evidence linking cholera cases to drinking water, skeptics continued to cling to the airborne explanation. One physician wrote that some of Snow's evidence was "most imperfect" and "quite worthless," concluding that for many of the patient examples he cited, "we do not think any one can feel that even a tolerable case is made out in favour of Dr. Snow's opinion." An editorial in *The Lancet*, a medical journal, derided his research as a "hobby"; with this theory, "he has fallen down through a gully-hole and has never since been able to get out again."

About a decade later, scientific opinion turned in Snow's favor, and the waterborne theory was vindicated. In the late 1800s, the idea that microorganisms could cause diseases also gained traction. But scientists eventually became reluctant to accept that *any* disease could be passed through the air, and "the pendulum swung too far" in the other direction, one team of researchers wrote in a historical analysis. A key figure was Charles Chapin, an esteemed epidemiologist who concluded in the early 1900s that airborne transmission generally didn't play an important role in the spread of common contagious illnesses. "It will be a great relief to most persons," Chapin wrote, "to be freed from the specter of infected air."

Washington

A year had gone by. It was February 25, 2021; we were waiting for the vaccines to become available to the general public. The number of SARS-CoV-2 sequences in the global database had now passed five hundred thousand.

I looked up the latest tree on Nextstrain and searched for the names of scientists I knew. The software displayed only a small subset of the half-million available sequences—including all of them would be too computationally cumbersome—but I was still able to find samples submitted by Placide, from a thirty-nine-year-old man in Lualaba and a fifty-three-year-old woman in Kinshasa. They had been collected shortly after I'd spoken to Amuri; the team must have gotten the reagents, I thought. I found Victoria, with samples from Montevideo, Uruguay; and Nouar, with samples from Ramallah and Majdal Bani Fadil in the West Bank. I found Paraic's team, which had contributed sequences from La Crosse and Adams counties in Wisconsin back in October 2020, shortly after college students had returned to campus. I found many posted by Jemma's team from all over New Zealand: Auckland, Wellington, Canterbury.

When I hovered my mouse over dots on the branches to see the sample records, more and more scientists' names emerged. There was Tze Minn Mak in Singapore, Luigi Atripaldi in Italy, Isa Nuryana in Indonesia, Francisco Duarte-Martínez in Costa Rica, Yvan Butera in Rwanda, and on and on and on. An author list included hundreds of other names—and those were only a small fraction of those who had contributed. These scientists had drawn the trees line by line, branch by branch.

When I shared what I had learned about scientists' COVID-19 analyses with friends, they wanted to know if the data had made a difference. Had the studies prompted politicians to push through stricter lockdowns, travel closures, mask mandates? Was there a point to any of it?

Like my friends, I also wanted to see a clear, bright line from science to policy. Scientists found a connection between cases; therefore, the government made changes that saved lives. When Snow had presented his findings about the Broad Street pump to the parish's Board of Guardians, the effect had been immediate: "*In consequence of what I said*," he wrote (emphasis mine), "the handle of the pump was removed on the following day." Within a week or so, the neighborhood's cholera outbreak fizzled out. I wanted a story as simple as John Snow and the pump handle.

Sometimes those simple stories were true. The New Zealand team's investigation into the quarantine hotel transmission helped set a cascade of changes into motion. Staff members upgraded from standard medical masks to N95 masks, air purifiers were deployed in the hallways, and nurses no longer tested adjacent rooms sequentially. Katarina and Gage found that a large cluster of cases at a daycare were genetically identical or near-identical, suggesting that those people had transmitted the virus to one another; the daycare was temporarily shut down.

But sometimes the science seemed to disappear into a void—or even if policy shifted, the decisions weren't driven by the release of new research. When I asked Paraic if his study on college outbreaks had affected public health policy, he said, "I don't think anything changed." By the time his results came out in October 2020, the university where his lab was based had already taken measures to quell rising cases, such as conducting extensive testing, issuing shelter-in-place restrictions, and suspending in-person classes. At that point, much of the damage to the community had been done.

When I asked why the work was still valuable, he told me that the knowledge could help us respond better to future pandemics. And developing the technical procedures to track outbreaks now meant that next time, public health officials could more quickly determine whether policies were too lax. Scientists were playing a long game, one that I didn't really want to admit the possibility of needing; once this was over, I thought, surely we'd move on and never have to deal with a crisis like this again. (I had the same feeling of unease when later reading Dave Eggers's novel *The Every*, which was set slightly in the future and casually referred to these types of events in plural, as in "After the pandemics…") Because nothing like this had ever happened in my lifetime, I'd let myself believe that COVID-19 was the whole story. But it was only one chapter in a much longer story.

England

As I learned more about John Snow, I realized that the Broad Street anecdote wasn't as simple as I'd first assumed. Removing the pump handle wasn't the decisive factor that stopped cholera in its tracks; by the time that was done, many people had fled the neighborhood, and the epidemic was already sputtering out. In his book *The Ghost Map: The Story of London's Most Terrifying Epidemic–And How It Changed Science, Cities, and the Modern World*, writer Steven Johnson argued that "the *decision* to remove the pump handle turned out to be more significant than the short-term effects of that decision." The incident was important, he wrote, "because for the first time a public institution had made an informed intervention into a cholera outbreak based on a scientifically sound theory of the disease."

The ripple effects of shutting down the pump were more

complicated. First, the decision may have prevented another outbreak from erupting later. The wave of cases had likely started when a woman dumped pails of water, which she had used to soak her sick baby's soiled diapers, into a cesspool that leaked into the well. On the day the pump handle was removed, her husband also got cholera, and the cesspool probably became contaminated again; if the pump had still been operational, it could have sparked a new conflagration of disease.

This chain of events had been uncovered by a curate named Henry Whitehead, who initially didn't believe Snow's claim that pump water had caused the outbreak. Whitehead was confident that he could disprove the theory by interviewing surviving residents, but the data he gathered eventually convinced him—"slowly and I may add reluctantly," he wrote—that Snow was right. His detailed investigation helped convince a local committee that the outbreak was at least "in some manner attributable" to the impure well water, a small but significant step in the slow process of pushing the waterborne theory toward mainstream scientific acceptance.

Like Paraic, the cancer biologist in Wisconsin, Snow had a long game in mind. He told Whitehead, in what the curate described as a "calm prophetic" manner, that eventually, perhaps after they were both dead, "the time will arrive when great outbreaks of cholera will be things of the past; and it is the knowledge of the way in which the disease is propagated which will cause them to disappear."

I began to see the trees of SARS-CoV-2 sequences as a historical document, a way to say *This happened.* Even if many of the findings had come too late to make a difference for the more than two million people who had died from COVID-19 so far, it was, as Snow said, "the knowledge of the way in which the disease is propagated which will cause them to disappear." It was necessary to make sense of what had passed; to write it down; to record this chapter. To trust in the process of handing down stories. And to believe that others would read them, whether a decade or a century from now, and take them to heart.

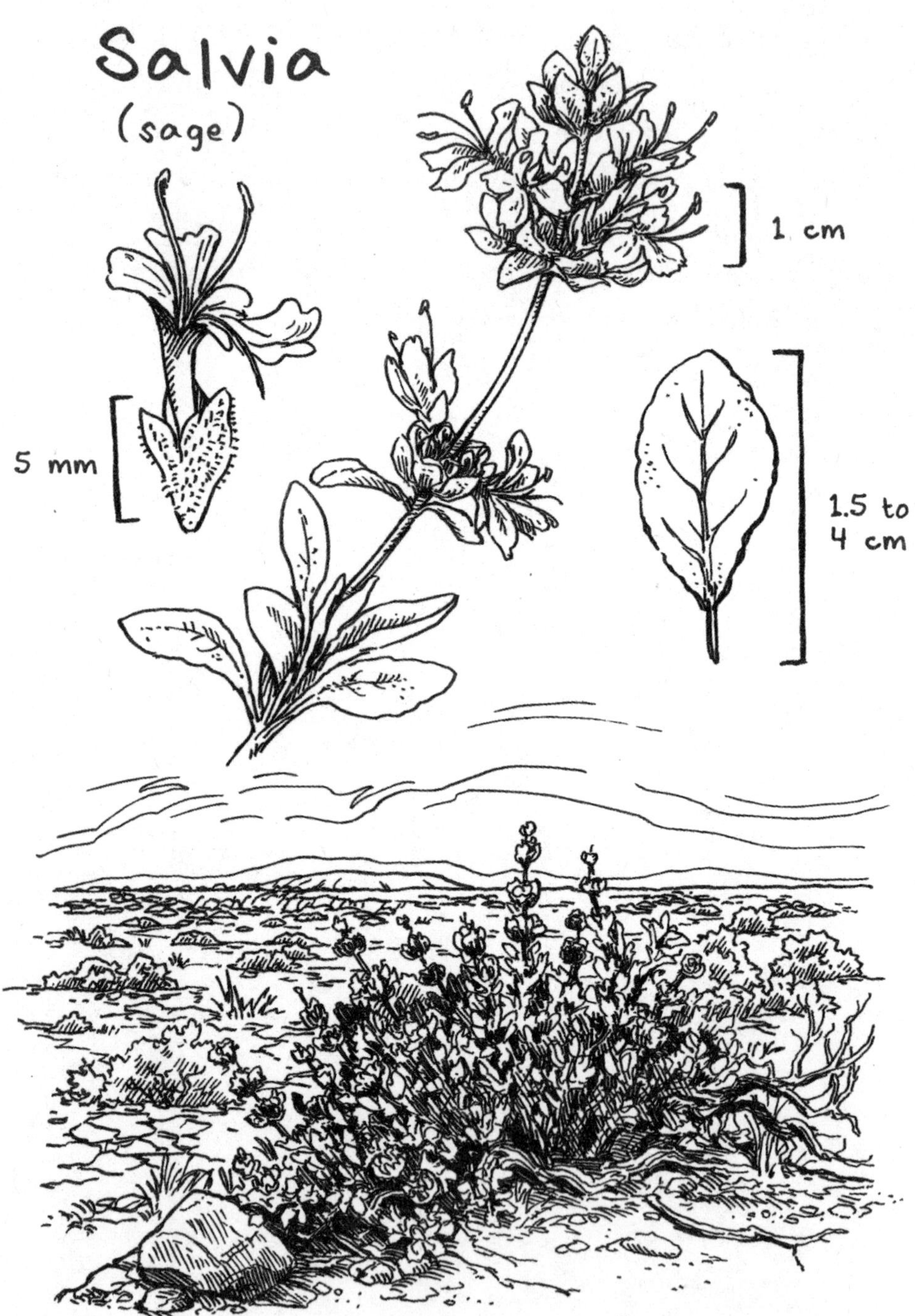

* grows in loose, sandy soil
in arid environments

Chemistry

A Space of Possibility

"Let us hope that reason and consciousness of right will exclude the wrong path."

On an April morning, I started driving from Tucson, Arizona, to Cameron, a small community about four-and-a-half hours to the north. I passed through the tangle of traffic in Phoenix, and then headed higher in elevation, through the mountain city of Flagstaff, past fields of pale yellow grass dotted with sparse bushes. At some point I entered the Navajo Nation, more than 25,000 square miles of land that cut across Arizona, New Mexico, and Utah, where at least 150,000 tribal members lived. Cameron was on the far west side, about a thirty-minute drive from Grand Canyon National Park.

The soil turned reddish brown, and I recognized the color. A couple of days earlier, I'd seen similar soil in plastic bins at the back of a greenhouse in Tucson. Richelle Lynn Thomas, an environmental toxicologist and PhD student at the University of Arizona, had taken me there to show her experiments in progress, using samples that she'd collected from the Navajo Nation. Richelle had crouched on the greenhouse floor and started scooping the soil with a trowel and shaking it through a sieve, red dust rising in puffs around her.

She was investigating the repercussions of intensive uranium mining that had taken place on the Navajo Nation for several decades, starting in the 1940s. Richelle, who is Diné, wanted to know whether medicinal plants were taking up heavy metals from the soil, which could pose a health risk to her fellow tribal members.[1] She planned to grow two plants, sage and Navajo tea. She'd spike the soil with known amounts of uranium or arsenic—another element present in the ground, often exposed to the surface and concentrated in mine waste—and measure how much they accumulated in different parts of the plant.

In the greenhouse, she showed me seedlings in white cylindrical containers, which she'd planted in the fine reddish soil and nestled in a black tray. The sage had stout bunches of thick, ruffled leaves; it could be used for smudging or in a lotion or tea and was meant to be cleansing and purifying, she told me. The Navajo tea resembled a spindly dark green grass, and the plant was usually folded into a bun and steeped in water to treat respiratory illnesses and cough.

Richelle was well-versed in their use, as her grandfather had

---

Some people in the tribe describe themselves as Navajo or as members of the Navajo Nation, while others prefer the term *Diné* (a word from their own language, which means "the people").

been a practitioner of traditional medicine and she and her sister had collected plants with him as kids. For Richelle's first graduate research project, she had decided to study the effects of uranium and other heavy metals in the environment. When she told her grandfather of her plans, he wanted to know how the metals affected people's health when they used medicinal plants. He'd passed away shortly afterward. "It's a question I always promised myself that I would answer for my grandpa, whether he was here or not," Richelle said. She saw her study as more than just a research project: "This is my life project."

Richelle's work was part of a larger story, a line of inquiry that stretched over more than two decades. She had started her research as a master's degree project with Jani Ingram, a chemist at Northern Arizona University in Flagstaff and a member of the Navajo Nation, who had been investigating the effects of uranium contamination since the early 2000s. Before joining the university, Jani had traveled to Shiprock, New Mexico, to collect water and soil samples as part of a U.S. Department of Energy project on radioactive waste management. When she arrived, she realized that the old uranium mine waste site was only about half a football field away from people's homes. She thought, "We need to do something about this."

Jani grew up in Kingman, a small town in western Arizona. She was a jock who liked math; she thought she could be a linebacker for a football team, a sports statistician, a bank teller. Growing up, Jani had absolutely no thought of being a chemist. But that changed when she took a chemistry class in community college with a professor who was approachable, encouraging, and—because he was married to a Navajo woman—understood her culture. During a summer internship and undergraduate research project, she realized that she liked making measurements, using math to get the answers, examining things at the level of atoms and molecules.

Jani had a few relatives who had worked in the uranium mines, but she hadn't heard much about the aftermath. When she was growing up, the community wasn't yet fully aware of the health repercussions. "People weren't afraid of it," she said.

Uranium mining had taken off in the southwestern United States during the early years of the Cold War, when the federal government was keen to establish a reliable domestic source of the metal for nuclear weapons. Large deposits were available on the Colorado Plateau, an expanse of Arizona, Utah, Colorado, and New Mexico that overlapped with the Navajo Nation. With the Atomic Energy Commission's encouragement, companies started underground and open-pit mining operations; for the next couple of decades, the U.S. government was the sole allowed buyer of uranium ore in the country.

For Diné men, the mines seemed like an economic opportunity—a way to earn income without having to travel to faraway jobs on the railroads. At the time, the government was already aware of reports that uranium miners in Germany and Czechoslovakia had suffered from high rates of lung cancer. But many Navajo miners said in later interviews that they received no warnings about the risks. In the 1950s, the U.S. Public Health Service even performed a study to better understand the health hazards of radiation exposure in uranium mines, without telling participants the full purpose of the study. Perry H. Charley, a professor emeritus who is Diné and worked for Diné College in Shiprock, saw his father and other miners line up, stripped to the waist, waiting to be examined in a green van. The government was "virtually certain" that these miners "faced a death watch," Charley said, but feared that knowledge of the risks would trigger a mass exodus. "Our fathers, our uncles, our grandfathers—they were never told of the dangers," he said. "They were just left to die."

Esther Yazzie-Lewis, a Navajo linguist and board member of the Southwest Research and Information Center, who is based in Edgewood, New Mexico, translated many interviews with miners. She said that they told her: "We didn't know it was hazardous. We didn't know anything about it. We went down to the mine, we took our lunch boxes down to the mine, and here we are sitting in uranium dust… We ate our food down there. We saw water coming out of the rocks, and that water tastes good. We didn't know that water had uranium in it."

Several of the students who eventually joined Jani's lab were Diné and had a family connection to the mines. One of them was Lydia Edgewater, whose grandfather had worked as a miner for about six months; people had told him that it was good money, and you didn't have to go far. But he said, "'I just didn't have a good feeling about it,'" recalled Lydia, who is now a scientist at the materials science company W. L. Gore & Associates in Flagstaff. He'd said, "'I just wanted to leave.'"

By the early 1950s, scientists had figured out how uranium causes cancer. Uranium is a big, heavy atom that "decays" over time; it transforms into a series of other elements and releases radiation at each step. For the most common type of uranium, one of those intermediate elements is radon, a gas; radon then decays into atoms called radon "progeny." Those progeny could latch onto dust particles, get stuck in the lungs, blast the tissue with radiation, and cause DNA mutations. Ventilation in the mines started improving in the 1960s, but even the reduced exposure would be considered unacceptable by today's standards, said Doug Brugge, an environmental health researcher at the University of Connecticut in Farmington.

It wasn't just the miners who were exposed. Their families often traveled with them and lived nearby; men came home covered in dust or mud, and their wives washed their clothes with the rest of the laundry. Children played on tailings, the piles of leftover crushed ore. And sometimes the miners brought back canisters of water for their families to drink.

In the 1960s and 1970s, many Diné miners started getting sick; in particular, community members noticed a much higher incidence of lung cancer. Wives and children saw that "all our men folks are dying off," Yazzie-Lewis said. Some Indian Health Service physicians were adamant that the disease was due to mining, not smoking. Few Diné men were heavy cigarette smokers, and they used only small amounts of tobacco for ceremonial purposes.

Tommy Rock, a former PhD student in Jani's lab who is a member of the Utah Navajo, recalled riding with his grandfather to a uranium mine where he had worked. His grandfather was normally chatty and liked to reminisce about the good old days. But at the mine, his grandfather stood quietly for a long time, staring at a large rock at the entrance. Tommy didn't find out until later that his grandfather's friend had been killed there, crushed by the boulder.

His grandfather, a World War II veteran, had always been strong and worked hard. But he suffered from asthma, eventually had to use a wheelchair with an oxygen tank, and died from cancer. When he was near death, "he fought to the very end," wrote Tommy, now an environmental scientist and assistant professor at Northern Arizona University, in a 2019 article. His relatives "told him it was okay to go, and that we would take care of our grandmother. He finally let go and left this world."

As I drove down Highway 89 through Cameron, I passed a truck stop, a couple of gas stations, and a Burger King, with smatterings of houses in the distance. In 1966, a federal government official had "frozen" nearly all development on about 2,500 square miles of land on the Navajo Nation; even after the freeze was lifted several decades later, many people still lacked basic services such as electricity and potable running water. Farther down the road was the Little Colorado River, spanned by a bridge with orange columns, and the Cameron Trading Post, a motel and gift shop near the river gorge with a steady stream of tourists. I planned to meet Richelle and three of her colleagues the following day; in addition to conducting her greenhouse research, Richelle wanted to help the community by screening soil samples from people's farms and yards for uranium, arsenic, and lead.

The next morning, I headed to the Cameron Chapter House, a pale blue, one-story building in a large gravel lot off the highway. The Navajo Nation is a sovereign nation with its own system of government, and the chapter house is a local unit where community members can bring up problems. A sign proclaimed "Administration Office" with a red arrow; the area was quiet except for the hum of passing cars. Inside, tables and chairs were arranged under fluorescent lights, and public health flyers were stacked to the side. One poster about a radiation exposure clinic offered cancer screening, and a map showed abandoned uranium mine sites on the west side of the Navajo Nation, marked with colored dots. A spray of dots clustered around Cameron.

When uranium mines associated with the nuclear weapons program were shut down, the companies "simply packed up and left," said Chris Shuey, an environmental health scientist at Southwest Research and Information Center in Albuquerque. "They were under no requirements to do anything to manage, control, consolidate, cover those wastes." More than five hundred mines were left behind on or near the Navajo Nation, as well as more than one thousand mine "features," such as sites where ore had been dumped. Some waste piles were abandoned right in the middle of communities, with no fences or warnings; families who were unaware of the danger used the leftover rock from mines to build homes. People who didn't live near a mine site could still be exposed to contaminated dust blown by wind or waste carried by rivers and creeks.

Outside the chapter house, Richelle's team had set up a registration table with packets of free seeds for tomatillos, cilantro, Navajo copper popcorn, snap peas, and Hopi casaba melon. Richelle was greeting community members, appearing cool and collected in a heavy-looking, velvety black shirt despite the heat of the day. Around the corner of the building, one of her advisers, environmental geochemist Robert Root, analyzed samples under a white canopy. Robert held a chunky black-and-yellow instrument resembling a squared-off hairdryer, pressed the end against each baggie of soil, and read numbers off a screen. He'd explained to me earlier that the instrument shot X-rays into the sample and kicked out an electron near an atom's nucleus; another electron then dropped in to fill the "hole," causing a photon to be released. By measuring the energy of photons hitting a detector, they could determine how much of each element was present.

One of the first community members to show up was Lula Neztsosie, a member of the Western Navajo Farm Board. Her father had worked in the mines and hadn't gotten cancer. "That word never crossed our mind," she said. But when she and her brother were young, they'd played in the rain-filled holes at mine tailings sites to cool off on hot days, and her brother had later been diagnosed with stomach cancer.

After advocates pushed for compensation for two decades, Congress passed a law in 1990 to provide "partial restitution"—a payment of $100,000—to people who had worked in the mines from 1947 to 1971 and later developed health problems, or to their surviving family members. But compensation was a "very stringent, stingy kind of process," Brugge said. The level of radiation exposure required to qualify was high and didn't account for the possibility of measurement errors. Some men who had smoked only for ceremonial purposes were still classified as smokers and needed an even higher level of exposure to receive compensation. Widows had to provide documentation to prove their relationships to the miners, but Diné people sometimes didn't have state marriage licenses.

The act was amended a decade later to lower the exposure threshold and remove the smoking provision, among other changes; in 2025, Congress expanded eligibility for miners again. But to date, no compensation has been provided to people whose families were not directly involved with the mines but live close to abandoned sites. The argument is that there isn't enough data to show that past mining affects the health of people here today, said Candis Yazzie, vice president of the Cameron Chapter, who is Diné. "But we're living it and we know."

In the late 1990s, doctors started noticing unusual levels of kidney disease on the Navajo Nation. The disease was more severe, progressed more rapidly, and started earlier than in other Indigenous communities. "It wasn't uncommon even at that time to see teenagers who had dialysis ports in their arms," said Johnnye Lewis, an inhalation toxicologist at the University of New Mexico in Albuquerque. Lewis's team found that more exposure to uranium mines, mills, and waste sites was linked to an increased risk of high blood pressure; those people also had a higher likelihood of multiple chronic diseases, such as kidney disease and diabetes. A more recent study suggested that uranium exposure was associated with an early indicator of autoimmune disease development, a type of disorder in which the body attacks its own cells.

Research on children yielded worrisome results too. In a report published in 2022, Lewis's team studied fifty-two pregnant women living on the Navajo Nation and tested their babies immediately after delivery. More than two-thirds of the mothers had elevated uranium levels in their urine, likely from drinking contaminated water or breathing in tainted dust. Uranium was also detected in some of the newborns' urine, suggesting that the contaminant had passed from the mother to the baby.

Close to the Cameron Chapter House, and less than a mile from an old mine site, was a preschool. It was a squat beige building with green and pink play structures outside; Ray Yellowhorse, a general maintenance worker for the chapter, had brought in soil from the preschool to be tested by Richelle's team. The X-ray analysis showed that uranium, arsenic, and lead levels in the sample were within a normal range, but Ray had other concerns. He wondered if community members could be unknowingly exposed to uranium when they collected clay for medical purposes. "Medicine men don't carry Geiger counters," said Ray, who is Navajo. While he knew the locations of abandoned mine waste, natural outcroppings of uranium weren't marked as dangerous. "It's never brought up and never tested," he said.

As people brought in samples to be analyzed, Richelle gave a presentation about her research in the chapter house to about ten community members. She told them that she wanted to carry on her grandfather's legacy with this project. "I climbed mountains with my grandpa, I went into the springs with my grandpa, I went into the canyon with my grandpa," she said. "I crawled, pretty much, in the reservoirs, to get these plants for my grandpa." Richelle assured the people assembled in the chapter house that she understood the effects of overharvesting plants, and she took a Navajo approach to her fieldwork. Earlier, she'd told me that she'd gotten permission to perform the study from the Diné Hataałii Association, an organization that worked to protect traditional ceremonies, as well as approval from chapter officials during her earlier master's research. When she first went into the field as a student with a medicine man, a medicine woman, and an herbalist, they said, "You just don't take what's not yours." She needed to make an offering and say a prayer every time she gathered a sample. Richelle had seen other scientists collect specimens in the field and then "just take off," she said. They didn't stop and think, "Is this going to be respectful?"

Someone in the audience asked how they could get more science done in the community, and Candis, who had been watching the presentation, spoke up. She said, "We do it ourselves."

When Jani started her research on uranium contamination in the early 2000s, one primary concern was water. On the Navajo Nation, many people didn't have access to public or regulated water, so they hauled it from unregulated wells. "Especially in the Southwest, a desert environment, water is key," said Jonathan Credo, a former member of Jani's lab who is Navajo and Filipino. "Water is life." The team started collecting water samples from communities near Flagstaff and then expanded to about three hundred samples from sites across Arizona and Utah.

Finding the wells wasn't easy. Navigation was tricky, getting to water sources often required driving off-road through sand or mud, and their locations were sometimes closely guarded and known only through word of mouth. Jonathan and Tommy Rock slowly gained the trust of community members, who told the researchers, "'Here's where we get our water,'" Jonathan recalled. "'Do not tell anyone else.'" The team drove through long stretches of desert to reach some sites, and Jonathan climbed on top of the car with binoculars, searching for the slightest glint or movement of a windmill powering a pump in the distance.

Back at the lab, the team tested the water samples for twenty-one elements. They found that although some sites did have high levels of uranium, "there are other elements that are arguably a lot more concerning," said Jonathan, who is now a resident in internal medicine and psychiatry at the University of California, Davis. The American Southwest is naturally rich in arsenic, he noted, and the researchers found arsenic levels above the Environmental Protection Agency (EPA) drinking water standard in forty samples. They also reported high levels of manganese, an abundant element in the earth's crust that can have neurotoxic effects, in twenty-nine samples.

Other students in Jani's lab investigated the potential health risks from eating sheep, a traditional livestock animal. The researchers wanted to know if uranium levels were higher in sheep from mining areas, similar to the way that mercury might be elevated in certain fish. In one study, the team found that sheep from two towns on the Navajo Nation had higher concentrations of uranium than those from a farm outside its borders.

The purpose of the studies wasn't to shut down all the wells or stop people from eating sheep. The results are more about raising awareness, said Colleen Cooley, a Diné facilitator and educator from Blue Gap and Shą́ą́'tóhí on the Navajo Nation, who worked in Jani's lab as an undergraduate. "How they choose to use that information would be up to them," she said. "You can't tell people, 'Because of this, you can't drink this water. Because of this, you can't eat your sheep.' That's where they live, and that's where they graze their sheep. That's where their livelihood is." But based on the data, people could choose to limit how much they drank from a well or to buy a filtration system, Jonathan said. The sheep data "has been definitely helpful" to her family, Candis said. People who wanted to be cautious could, for instance, get their mutton from a less contaminated location, she said.

For Marissa Mares, a former master's student, working with

Jani meant that she could pursue her interests in both natural sciences and social justice. "My people are still dealing with this and they're still concerned about it," said Marissa, who is Diné. "It's truly a terrible legacy that's being carried on. The reason that I'm so passionate about it is because that I can see that it's so wrong."

The research in Jani's lab also helped counteract the usual narrative of being only victims, Lydia said. They were taking control and doing something. "We are the future, we can research, we can educate, we can learn more about this so that we can speak," she said. "So this won't happen again."

The day before we left for Cameron, Richelle took me to Herring Hall on the University of Arizona campus, a brick building with white columns and a hallowed feeling when we stepped inside. The front room had high ceilings, a long table with microscopes, and delicate illustrations of plants on the walls. The building was an herbarium, a plant library of sorts; down a flight of stairs toward the back, we entered a windowless room full of tall, green cabinets, which held plant collections sorted by family, genera, species, and region.

Richelle opened a cabinet to reveal drawers of large folders, each containing pressed specimens. One was devoted to Navajo tea, which belonged to the genus *Thelesperma*—long, spidery, dark stalks with small yellow flowers. Entries were dated 1945, 1963, 1971, 1986, 2015; the older ones appeared to have been written on a typewriter, the letters arranged crookedly on the page, while others had been scribbled in black ink.

From the folder for sage, called *Salvia*, Richelle took out a sheet for a 1992 sample that had been collected on the Navajo Nation. There were three plants with grayish, nubbly leaves and dark flowers that were once a vibrant bluish purple, a dried bouquet frozen on the page. She read the caption and said that this was the type of sage she'd grown up harvesting with her grandparents. "I always wish he was still here," she said of her grandfather.

Earlier that day, Richelle and I stood in front of a crinkled periodic table posted in the hallway of another building on campus, where one of her advisers had a lab. Uranium was in the bottom row, three boxes from the left. According to the poster, it had been discovered in 1789 and was named after the planet Uranus. The element was present in many deposits around the world; on the Colorado Plateau, it often came in the form of carnotite, a bright yellow mineral.

In an illustration on the periodic table, uranium was depicted as a shiny, perfectly formed silvery cylinder. It looked smooth, innocuous, a marvel of modern technology. In the late 1930s, scientists in Europe had discovered the process of nuclear fission, which would ultimately power the bombs dropped on Hiroshima and Nagasaki. If neutrons hit a certain type of uranium, the atom could split into different elements and release energy and more neutrons, setting off a chain reaction. Lise Meitner, an Austrian physicist who was involved in the research, later wrote about the development of nuclear weapons: "Let us hope that reason and consciousness of right will exclude the wrong path. But there's an old proverb: It's later than you think."

After Richelle's team finished the soil screening in Cameron, I drove along a dirt road leading away from the chapter house, looking for one of the abandoned uranium mines. I passed houses, trucks, and a couple of horses, and then turned around and saw a sign with a yellow-and-black radiation warning symbol that said "KEEP OUT." The mine site was a hill of gravelly, pale yellow rock, with scattered shrubs and small trees. A few tires had been left at the edge, along with wooden pallets, a rusted cart, and a pink toy motorcycle that had faded in the sun. Long grass grew between some of the rocks. Driving along the road, I saw sign after sign; they warned "No building, gathering, playing, corrals, or digging at mines." The houses, the trailers where people lived, and the preschool were so close.

Government workers were trying to get some of the mine sites cleaned up, but they were nowhere near done. In 2007, Congressman Henry Waxman had ordered the EPA and other agencies to tackle the contamination and community health problems more seriously. Under the so-called Superfund law, the EPA could sue the mining companies or their successor firms and get money through settlements to pay for remediation; or the agency could order the companies to do the job themselves. As of this writing, agreements and settlements were in place to conduct cleanup operations estimated at more than $1.7 billion; the EPA had assessed more than one hundred sites and performed urgent tasks at about thirty locations, but only one site had been fully remediated.

For many community members, progress felt frustratingly slow. "They're working, but it's like a snail's pace," Yazzie-Lewis said. Will Duncan, assistant director in a Superfund and emergency management division at the EPA in San Francisco, didn't dispute that interpretation. "I don't fault anybody who thinks that things have moved slowly," he said. "I think they have." But the process was deliberately long to make sure that the agency addressed all the contamination, he said. Investigating sites was challenging, partly because waste was often spread over a large area.

Duncan estimated that the settlements would address roughly two-thirds of the waste on the Navajo Nation, and he said that agencies had prioritized the "worst of the worst" sites to clean up first. In early 2025, the EPA announced a plan to move more than one million cubic yards of waste from one site, Quivira Mines, to a landfill repository in New Mexico.

Many mines were "orphaned" without anyone to hold accountable—because the companies either had gone out of business or didn't have enough money. The federal government should take responsibility for remediation of those sites, argued Michelle David, now a judicial law clerk at the U.S. District Court for the Northern District of Georgia in Atlanta, who wrote an analysis of the issue. The government had facilitated the mining, developed contracts, steered operational decisions, and controlled the market. And it had been the sole procurer for many years, Charley said: "Ultimately, they were the masterminds."

The federal government is paying for cleanup at sixteen high-priority orphan mines and assessment of another thirty orphan sites, as well as partly covering remediation costs at other abandoned mines. But in Shuey's view, those actions still fall far short of what's needed. The government channeled the equivalent of more than thirty billion of today's dollars into the Manhattan Project to produce nuclear bombs for World War II. And yet "we do not have that same effort to clean up abandoned uranium mines, the vast majority of which were created for the nuclear weapons program," Shuey said. To remediate the mine sites, "we need a Manhattan Project–like effort, expertise, funding, statutory and regulatory authorities, at the same level of zeal and zealotry."

When Shuey said this to me, I felt the absolute truth and reason of his argument, but at the same time I felt certain that government leaders would never do this, would never spend $30 billion to clean up their own mess, would never fully take responsibility for what had been done, no matter how many sad stories they heard. The Navajo people are "bowing and saying 'Come on America, help us, help us,'" Yazzie-Lewis said. "It's still messed up. People still cry."

After I drove back to the chapter house, I thought about something I'd seen when I stopped for lunch at the Burger King in town. A young man in an employee uniform, who looked Diné, had come outside to empty a trash can and put a fresh bag in the bin. A woman, whom I pegged as a likely tourist based on her summery clothes and carefree attitude, walked up to him and tossed her trash into the full bag he was holding and then walked away eating her ice cream sundae. Something about the look on her face bothered me. I knew I was making assumptions; he might not have been Diné, she might not have been white. Still, I couldn't help seeing them as stand-ins for a larger story.

About a month earlier, I'd been talking to a fellow writer about sustainability, and I asked if her parents were environmentally minded. She said no, they were like the scene in *Mad Men*, where the Draper family has a picnic in the park and then tosses their garbage on the grass and leaves. As I recalled, she described their attitude as "The world is for us." That was what the look on the woman's face at the Burger King reminded me of.

The contamination left behind on the Navajo Nation hadn't stopped another company from trying to extract uranium from the region again. NuFuels, a subsidiary of the Canadian firm Laramide Resources, planned to mine uranium near Church Rock, New Mexico, and process it in nearby Crownpoint. The proposed sites were in the "checkerboard" area, a mishmash of Navajo, federal, state, and privately owned land, as well as parcels allotted to individual Navajo people. In 2023, Laramide announced that it had performed initial drilling; the following year, the company released the results of an economic assessment and concluded that mining the area could be a "long life, high margin project."

The firm planned to use a newer technique called in situ leach or in situ recovery mining, which would involve releasing uranium from deposits in an aquifer that supplies drinking water to Diné community members. If these deposits are left undisturbed, the drinking water in other parts of the aquifer can still be high-quality, said Eric Jantz, legal director at the New Mexico Environmental Law Center in Albuquerque. But the process of extracting uranium "can contaminate huge areas of aquifer," he said. Jantz was skeptical of claims that the groundwater could be returned to previous conditions. Other mines around the world have used the same technique, "and not a single one has been able to restore groundwater to pre-mining quality," he said.

When I contacted Laramide, the company referred my questions to Janet Lee-Sheriff, president of the Clean Energy

Association of New Mexico in Santa Fe. Sheriff said that the water was already contaminated by uranium and shouldn't be used for drinking. The project would also be required to include monitor wells to detect leaks. Regarding community members' concerns, "I completely understand why they would never want to see conventional mining," she said. "This, to me, is not mining. This is an environmentally responsible means of extracting something that you don't want around and removing it."

The New Mexico Environmental Law Center and a nonprofit representing community members, Eastern Navajo Diné Against Uranium Mining, were trying to stop the project. After challenging the license and appealing to the courts failed, their latest strategy was a petition to the Inter-American Commission on Human Rights (IACHR) in Washington, DC; as of this writing, the organization was still deciding on the case's merits. Even if the IACHR ruled in favor of the petitioners, it would have no real power to stop the mining.

I cast my mind back to a more hopeful scene, the second day of the soil screening in Cameron, when a man named Nolan Stevens had dropped off samples. He had a kind-looking face, with a sparse mustache and slicked-back black hair with gray streaks in a ponytail, wearing a blue bandanna and mirrored sunglasses on top of his head. He looked strong, which made sense when he told me that he'd spent fourteen years working at a gold mine in Nevada. Nolan, who is Diné, had big plans: For more than a decade, he'd been saving his money to start farming in the Cameron area, where he'd spent part of his childhood, and to bring some of his kids and grandkids back to the Navajo Nation. They'd had a good life in Nevada, but his children were missing their identity. He'd started buying equipment, and he wanted to hire workers and improve the local economy.

Nolan had brought soil samples from two farm sites to be tested, and the results came back as low-risk. He was relieved. "I feel good about these readings," he said. "It's a load off my shoulders. Especially with what I had riding on it."

During the soil screening, Richelle's team had visited a hoop house nearby, a tentlike structure with a semicircular roof and a translucent plastic covering over a metal frame. Shelves of gardening supplies, stacks of planting trays, and white tarps were scattered along the edges. Richelle had looked around, assessing the options, and said they could use half of it for planting food and the other half for Navajo tea and native plants. She also wanted to set up a greenhouse in nearby Tuba City as a place where people could learn about the results of her experiments, safe gardening practices, and how to grow traditional foods. The hoop house was light and bright inside; it felt like a space of possibility.

I thought of something that Perry Charley had said about the legacy of uranium mining. He told me that, even after decades of investigating the impacts, "I never saw the end, of all of this." It didn't seem likely that there would ever truly be an end—a satisfying and fair and just end—no matter what scientists found and how thoroughly they documented the damage. But hearing Nolan's dreams and Richelle's plans, it felt like, in spite of everything that had been done to the land and the people living on it, in spite of that overwhelming narrative of destruction and disregard and moral failure, there was still space for small new stories to take hold. Instead of a mine, a farm. Instead of a waste pile, a garden. Instead of taking and leaving, staying and growing.

Fossil Tusk
Cross-section
(magnified view)
Lystrosaurus
Tusk Fossil
UWBM
109750

PALEONTOLOGY

Through a Glass Darkly

"It just finally made sense to me."

Lystrosaurus did not seem like the hero of any story.

A stumpy, four-legged animal that lived roughly 250 million years ago, it resembled a cross between a warthog and small hippo. The creature had a squarish head, fleshy midsection, downturned snout, and short tail, with forelimbs that splayed to the sides and two tusks. When I called paleontologists to learn more about *Lystrosaurus* (pronounced "LISS-trah-SOR-us"), they spoke about it with affectionate humor, like parents whose child had turned out a little weird. "He was a strange-looking thing," said one scientist. And with its sprawling front legs, "he would have walked quite funny as well." Another researcher, attempting to find words to describe the animals, said, "I love them, but…"

and then added, "we're not talking about, you know, beautiful cheetah in motion."

I had first learned about these creatures from Meg Whitney, then a paleontology PhD student at the University of Washington in Seattle. She called *Lystrosaurus* and related animals "derpy" because of their sad, long-faced appearance. Meg planned to examine slices of *Lystrosaurus* tusk fossils under a microscope, which would allow her to deduce certain facts about the animals' lives. Just as tree trunks grow in rings, the tusks grew layers over time, leaving behind concentric circles. "You can kind of read it like a storybook," she said.

Appearances aside, *Lystrosaurus* deserves our respect: These animals thrived while many other creatures perished in the deadliest mass extinction in Earth's history. Around 252 million years ago, volcanoes in Siberia started erupting and burbling. Carbon dioxide and other gases spewed into the atmosphere, setting in motion a series of no good, very bad events: The planet warmed, acid rain fell, and the ozone layer was probably punctured, letting in harmful ultraviolet radiation. In the oceans, oxygen levels dropped, and microbes churned out poisonous hydrogen sulfide. By some estimates, 81 to 96 percent of marine species and about 90 percent of tetrapod species—vertebrate animals with limbs—vanished. But somehow, *Lystrosaurus* made it through the madness. While other species disappeared, these piglike creatures waddled on.

Understanding *Lystrosaurus*'s history could help scientists predict which animals will survive the next wave of extinctions caused by climate change and habitat destruction, said Ken Angielczyk, curator of paleomammalogy at the Field Museum in

Chicago. Perhaps creatures with similar traits will weather today's slow-motion train wreck better. If certain animals can respond flexibly to stressful conditions, as *Lystrosaurus* did, "they're probably going to be more capable of dealing with whatever's coming down the pipeline," Meg said.

In January 2019, I visited Meg's lab in a glassed-in research building surrounded by evergreens, a medicinal herb garden, and a busy bike trail. When I arrived, she showed me a small white box. Resting inside, like a piece of ancient jewelry, was the weathered tip of a tusk. Someone had written "UWBM 109750" on its knobbly surface, and an accompanying pink slip listed the original location as "VP106.47." Though I didn't know what the labels meant, I felt compelled to write them down, like codes to be later unlocked.

"Is that the thing?" I said. "Can I touch it?"

"Absolutely," Meg said. This fossil, she said, was from Antarctica. I picked it up and waited for a feeling of awe, but the tusk felt no different from an ordinary rock. It did not quite sink in that this "thing" was hundreds of millions of years old, an expanse of time so enormous that it seemed meaningless. Neither could I fully comprehend that the tusk had adorned an animal that once roamed the land in huge numbers, and that this remnant had been sawed out of an ice-bound mountain on the other side of the planet. For now, it just felt like a rock.

The story that Meg wanted to read in the fossils was a prologue to human history. The tusks' growth patterns, she thought, might suggest whether *Lystrosaurus* was warm-blooded—a trait that could have made the animals less sensitive to environmental changes and helped them avoid extinction. Knowing that would help scientists narrow down when warm-bloodedness arose during the evolutionary path to people, which no one had quite figured out.

Warm-bloodedness, in many ways, makes us who we are. Some benefits are relatively modest: the ability to, say, endure Midwest winters without having to spend long periods sleeping underground. Mammals' ability to lactate and feed their young from an internal source might not be possible without this trait, Meg said. Christian Kammerer, a paleontologist at the North Carolina Museum of Natural Sciences in Raleigh, also noted that being warm-blooded enables us to perform intense physical activities. So if humans were cold-blooded, I asked, would we spend more time sitting around watching TV instead of running marathons?

"We may not even be able to understand TV," he said. Our brains would chug along more slowly, he speculated: "I don't think we'd have cities or real culture." Kammerer pointed to the naked mole-rat, a cold-blooded mammal that congregates in lumpy piles underground. We'd become "these hairless, sluglike, little sausage guys that just have to huddle together in pits to stay warm," he said—rather than, say, painting watercolors, designing skyscrapers, or pondering the origins of the universe.

Untangling when warm-bloodedness evolved required studying early creatures on the cusp of this transition. And *Lystrosaurus* was one of those old relations, a sort of distant cousin. It belonged to a group of animals called therapsids, which dominated the land before dinosaurs showed up. Some therapsids eventually evolved into mammals, so Kammerer liked to call them "proto-mammals." "Our ancestors ruled the Earth before the dinosaurs ever evolved," he said. They ranged from mouse- to elephant-sized. Mole-like creatures burrowed underground, a leopard-sized predator flashed serrated fangs, and a long-fingered animal with a monkeyish tail clung to trees. In the subgroup that evolved into mammals, the species included ten-foot-long, lumbering, likely furry animals resembling the R.O.U.S.'s (Rodents of Unusual Size) in *The Princess Bride*.

Among this zoo, *Lystrosaurus* had two terrific advantages for Meg. First, its tusks offered a handy way to study its life history. She and Athena Tse, a lab technician, had started counting the concentric rings in the fossils. Each band captured a short period of growth. If the animal had gotten stressed and stopped growing, a line would appear thicker. By analyzing the stress marks, Meg could determine how these creatures had responded to the extreme seasons in Antarctica and thus whether they were warm-blooded.

The second advantage: *Lystrosaurus* was everywhere. Scientists had found its remains in places such as South Africa, India, China, Russia, and Antarctica. If all the animals on Earth today were buried very suddenly, and curious scientists poked through the fossils millions of years from now, they would find zillions of squirrels; that's what *Lystrosaurus* is like for today's paleontologists. A hardy animal, undeterred by catastrophic planetary upheaval, frozen in time.

Meg and her labmates had recently moved into a space on the bright, airy second floor of the university's life sciences building. This was unusual, she later said, because paleontologists typically ended up below ground level. "People think 'rocks, basement,'" she said.

The lab had drawers and drawers of fossils: turtles, *Tyrannosaurus*, ancient cowlike creatures. The office section with desks and computers looked not too different from a Silicon Valley startup—blobby chairs in the coffee area, a jokey plastic severed hand perched on a doorknob—aside from the fact that the whiteboard displayed evolutionary trees and the skulls of ancient animals grinned from shelves. One hallway was lined with metal shelves holding dozens of large Ziploc bags of dirt. "This is all from Montana," Meg said later, with an exasperated look. In the field, it would have taken too long for the researchers to search the soil for small fossils. So they had brought it back to sort out later. "It's so much dirt though," she said. "It's *so* much dirt."

The tusk that Meg was analyzing today had been discovered in Antarctica a few decades ago by paleontologist Bill Hammer, who had traveled to the continent eight times and amassed an impressive fossil collection. Hammer, now a professor emeritus at Augustana College in Rock Island, Illinois, said that he had neglected some of his discoveries. The press got excited if he excavated a dinosaur, but "you find something else, nobody cares." So he had given the fossils to the University of Washington.

Meg's first task was to make a mold to capture the fossil's shape, just in case another scientist later wanted to know what the tusk looked like before she sliced it. She laid the fossil flat on a lab bench, brought out a tub of LEGO bricks, and built a little fort around the tusk. Then she patted clay around the bottom half of the fossil to hold it in place; moldable silicone rubber would be poured on top.

"If I'm having a bad day, this is the best thing to do," she said. "It's really therapeutic." She added, "We also cut things up on a saw. That's if you're having a *really* bad day." Meg worked with a casual confidence, her ankles crossed under her. Though she was, at the time, twenty-eight years old, she wore an old-fashioned wristwatch because, as she explained later, "I hate being late. I like to know what time it is all the time."

Meg wanted to compare tusks from Antarctica and South Africa, which had once been attached as part of a supercontinent. One might expect Antarctic animals to be more stressed because of their harsher environment. (Though Antarctica was far warmer at the time than it is today, darkness had still enveloped the continent every winter; "it doesn't seem like that would be a great place to live for three months of the year," said Chris Sidor, Meg's adviser.) But if stress levels didn't seem higher in Antarctica, perhaps warm-bloodedness had made the animals less sensitive to environmental changes.

In the past, scientists had considered *Lystrosaurus* and its ilk cold-blooded and reptilian. But some tantalizing signs had emerged suggesting otherwise. Based on their bone structure, the animals seemed to have grown very quickly—an indication that they had a high metabolism, which is characteristic of creatures that can maintain their own body temperature. "My inkling," Meg said, "is that way, way, way back on the mammal line, *Lystrosaurus* was definitely warm-blooded."

Some scientists remained unconvinced. "No way," Hammer said. "It's too primitive." Jennifer Botha, director of the paleosciences center GENUS at the University of the Witwatersrand in Johannesburg, South Africa, said that *Lystrosaurus* was missing telltale anatomical features. For instance, mammals have muscular diaphragms, which allow the animals to breathe more frequently, get more oxygen, and thus fuel the energy-intensive process of maintaining their body temperature. But *Lystrosaurus*'s barrel-shaped body didn't leave much room for such a structure. Nor did it have clear evidence of special membrane-covered bones in its snout, features that helped prevent too much water loss while breathing out, she said. In Botha's view, it was "highly unlikely" that the creature was warm-blooded.

But it wasn't an either-or scenario, Kammerer noted. Some animals fell along a continuum from warm-blooded to cold-blooded, rather than being at one extreme. For example, swordfish warm up their brains and eyes while other body parts stay cooler. Leatherback sea turtles are reptiles but can maintain a higher body temperature in cold waters. Therapsids, he suspected, landed in this middling zone.

Meg's hunch was that some therapsids were cold-blooded and some warm-blooded. These creatures were "incredibly diverse," she said. "There's probably a lot of evolution happening within each one of these groups that we underappreciate." That, she said lightly, was "my hill I'm going to die on."

Growing up in Maine, Meg thought she'd become a doctor. A high school biology class sparked her interest, and she was inspired by the TV show *Grey's Anatomy*. But as an undergrad student at Macalester College in Saint Paul, Minnesota, Meg was drawn instead to a biodiversity and evolution course. Every Friday, the professor, vertebrate paleontologist Kristi Curry Rogers, presented an "organism of the week": eerily glowing bacteria, for example, or a pill bug–like creature that sucked blood from a fish's tongue and eventually replaced the organ entirely. For the final exam, students had to place an assortment of those species on the tree of life, and Meg liked fitting the creatures into a story. When she later asked Kristi for career advice, the professor suggested that Meg join her lab to analyze fossils.

Meg didn't like dinosaurs; she thought it was silly to study dead things. But Kristi persisted. "Do a research project with me," she said. "See how you like looking at bones all day."

Meg did like it. She analyzed fossils of a baby dinosaur from Madagascar and got hooked on the practice of cutting up bones to see growth patterns. Meg spent hours grinding down fossil slices to thin layers, holding them against a wheel coated in abrasive diamond dust that occasionally also took off a layer of her skin. She bandaged her bloody fingertips and kept going. The tedium and minor injuries were worth it to see an ancient animal's story revealed under the microscope; at the end of the day, she still felt happy. She had a knack for "connecting the dots across time," Kristi said. "You have to use all your powers of reasoning to triangulate." For instance, Meg had scoured papers on everything from crocodilians to ostriches to figure out the meaning of a strange line they'd seen in the dinosaur fossils. She determined that it was likely a hatching line, a subtle indication of stress after birth.

After joining Chris Sidor's lab, Meg wanted to go on one of his Antarctica expeditions and told him so right away. A couple years later, she got her chance: The team traveled there to search for animal fossils from the aftermath of the mass extinction.

They camped on Shackleton Glacier, in a sort of mini-city of tents for cooking, eating, sleeping, science gear, helicopter logistics, mechanical equipment, and medical care. Sometimes people found remnants of previous expeditions: Meg's team discovered a newspaper and packaged creamer left behind by scientists several decades ago, and on another trip, a colleague ate a piece of abandoned Christmas fruitcake, frozen since 1969.

Since the continent is covered in an ice sheet averaging more than a mile thick, there was only one place to find rocks: on the tops of mountains poking out of the ice. On days with good weather, Meg's team headed out in the helicopter to nearby peaks and excavated fossils with rock saws and other tools. Chris gave Meg the unenviable task of making protective coverings for the fossils in the field, which required mixing plaster with water and shaping wet bandages around the bones before they froze. She wore plastic gloves and played music to distract herself from the burning, achy sensation in her hands. "I was sacrificed because I was the grad student," she said. "Which is fine."

The landscape was like a painting with only a few colors: white, blue, gray, the occasional reddish rock. Meg could see mountains and rocks and sky, but no trees or shrubs. Signs of life were absent. She couldn't train her eyes on anything, and it was hard to tell how far away things were. Out there, she lost her sense of perspective.

A couple weeks after my first lab visit, Meg was ready to slice the *Lystrosaurus* tusk. In a small room that she called the main operations center, she cut a two-millimeter wafer off the fossil, now encased in resin, with a whirring circular blade. Next, she needed to make the slice even thinner so light could pass through. Like a potter at her wheel, Meg held the fossil wafer against a spinning abrasive disc. This step wouldn't take long, but while working in Kristi's lab, she had once spent about twenty-five hours grinding a large fossil. "It was the worst," she said. She had passed the time as she worked by watching *The Daily Show with Jon Stewart* on the computer.

In the microscope room, she checked whether the wafer was thin enough. The magnified image of the *Lystrosaurus* tusk slice on the computer screen looked like a thick brown ring, similar to a tree trunk. As Meg inspected it, I mentally rewound millions of years and tried to picture this animal. It often lived on floodplains; ferns, club mosses, and horsetails grew along Mississippi-sized rivers. *Lystrosaurus* chewed shrubs and dug for tubers and roots. It probably slept in a burrow. And it dodged predators such as saber-toothed carnivores.

When the Siberian volcano apocalypse hit, these homely creatures might have survived because they weren't picky about which plants they ate. And as a burrowing animal, they likely had an advantage. If there was a drought, "I can go in my burrow and kind of pass out for a few months," Angielczyk said, and "resume life when times are better." Plus, maybe they were already accustomed to dealing with low oxygen levels, just as a mole can handle stale air if an oblivious cow steps on its tunnel entrance.

On a typical day post-extinction, *Lystrosaurus* would have emerged from its burrow, looked around the ash-blanketed landscape, and seen no living things other than thousands of other

*Lystrosaurus*es and a few ferret- or crocodile-like creatures—like the lone survivors in a sci-fi film roaming a deserted, bombed-out city, encountering only each other and the occasional cockroach.

Still, there were huge blank spaces in scientists' understanding of these creatures, waiting to be filled in. That didn't bother Meg. "I love how much we don't know," she said. She found meaning in thinking of humans as the product of incredible evolution over a vast time span: "We just get to experience it for a minute." During our conversations, she tossed around the terms "Permian" and "Triassic"—the periods from about 299 to 252 million years ago, and 252 to 201 million years ago, respectively—the same way someone would say "yesterday" or "last week." Journalist John McPhee described a similarly casual attitude toward enormous time scales among geologists in his book *Basin and Range*. "If you free yourself from the conventional reaction to a quantity like a million years, you free yourself a bit from the boundaries of human time," one scientist told him. "And then in a way you do not live at all, but in another way you live forever."

After some more grinding and polishing, the tusk slice was ready. "Yay!" Meg said.

"Is this the fun part?" I asked.

"I'm so excited," she said. We would be the first people to see the interior of this ancient fossil. "Seriously, it never gets old."

But when Meg got her first look at the tusk under the microscope, she sounded disappointed. "This is a pretty crap specimen," she said. "This is all a bummer." Long, spindly cracks splayed across the fossil cross section. It looked like a car windshield after an accident. The cracks made it harder for Meg to count continuous concentric rings, like reading a book with sections ripped out.

She moved the microscope view to a different part of the tusk. "That's nice," she said. "This is better." Meg decided to proceed because she had so few Antarctic specimens. "It's not going to be an awesome tusk," she said, but "it's going to have to make do."

Meg and Athena had devised a method to count lines more objectively, by measuring the light intensity at each point on the fossil. Then she observed where the light intensity dipped below a certain threshold—meaning that it was darker and, therefore, a line. But for this specimen, Meg had to rely on her eyes, because the cracks had broken up the lines.

To me, these lines were barely visible. "They're really, really faint," Meg said. "Which is why counting them by hand is a nightmare." And the subjectivity of the process bothered her. "I don't like that as a practice," she said. "Because there's no way to standardize this."

I asked if I could help. Still staring at the screen, she said, "*Ughhhhh*..." Then she suggested that I look at each line she identified and say whether I agreed that it was a line. Meg pointed at one: "I think that this one looks like a line."

"Y-e-e-e-s?" I said.

She laughed and pointed to another possible line. "I don't know if I see one at that exact spot," I said. "If I follow it up..."

"Maybe right there?"

"I don't really—maybe if you follow it that way?" I said tentatively.

"If in doubt, we're not going to call it. How about right there?"

"No..."

Another line. "Yes."

"Okay."

By the end of our session, she had gotten only partway through the tusk. But she had perked up again. "I was really pessimistic when we opened this up, but this looks pretty good actually," she said. "I think this is going to work."

Over the next month, the Seattle area was hit with more snow than the city had seen in seventy years. Our neighborhood transformed from damp greenery to a white expanse, like a sped-up film of Antarctica chilling from forests into ice; we entered a frozen diorama. When I returned to the lab and joined Meg in the microscope room, she pointed to a lighter area on the magnified tusk slice without lines, which suggested the animal had been growing fast.

"If I see a big white band like that, what that tells me is that the animal was kicking ass," she said. Then she pointed at the corresponding peak in a graph of light intensity: "This peak is happy happy fun times."

"It probably had a lot to eat?"

Meg agreed and translated my interpretation into scientist-speak: Maybe "it happened upon a good nutrient source." This part of the tusk, she said, captured the animal's last few months. It didn't appear very stressed, so perhaps it had been gobbled by a predator rather than suffering a long, drawn-out demise. "It's kind of why I like the histology stuff," Meg said as she mused about the cause of death. "It's a very intimate relationship with the animal."

Afterward, we went to her desk to look at a graph of her data. Next to the computer, Meg had arranged a LEGO set of female scientists; one figure with shoulder-length, wavy brown hair, who was holding a magnifying glass and standing next to a *Tyrannosaurus rex* skeleton, looked kind of like Meg. "She has a microscope, which I like," Meg said of her doppelgänger.

"She also has tusks," I said, pointing to a stack of pointy bones the figure was standing on.

"They're the backbone," she corrected me. "But I'm definitely going to make them tusks." She took a pair off the stack and attached them to the dinosaur's mouth like a jaunty mustache. "I can't believe I didn't think about that before."

So far, Meg and the lab technician had analyzed three South African and two Antarctic tusk slices. Meg had noticed very thick lines in the Antarctic specimens, which seemed to be signs of intense stress—but she couldn't yet say whether these bands argued against warm-bloodedness. She needed to add more data and run statistical tests. "Hopefully this new specimen will fall out somewhere interesting," she said. "It could go either way."

In his book *On the Origin of Species*, Darwin remarked on the gaps in the geological record, which he called "a history of the world imperfectly kept." He wrote, "of this history we possess the last volume alone… Of this volume, only here and there a short chapter has been preserved; and of each page, only here and there a few lines." Back at home, I opened a photo of the *Lystrosaurus* tusk slice on my computer. Here was the history book; I would try to read it. At first, I couldn't see the lines at all. If I squinted, some faint tracks appeared. Zooming in just made everything fuzzier; the closer I got, the less I could see. As soon as I tried to train my eye on a line, it disappeared.

Five months later, on a brilliant sunny July day, I returned to Meg's lab. She was in good spirits, despite being jet-lagged from a recent trip to Zambia to collect fossils. Meg had found a huge skull of an animal closely related to *Lystrosaurus*, just lying face-up on the side of a road, and "lots and lots and lots" of its smaller relatives. "Really cute little tiny skulls," she said. "They're rats basically. But they're cute rats."

She was nearing the end of her *Lystrosaurus* study, but she still needed to analyze more tusk specimens and double-check the data. So far, she said, the Antarctic and South African animals seemed to have similar growth patterns, which suggested that Antarctic creatures hadn't been much more stressed. For now, at least, the results favored warm-bloodedness. In the microscope room, I flipped through Meg's lab notebook; the technical descriptions meant little to me, but the words conveyed a struggle for precision, a resolve to make do without perfect data:

better preservation but not consistent
how measure line thickness? even on 20x is too small?
HAM 71 no lines; poor slide making
HAM 22 not great but countable

On one page, she had scribbled *in fossils it is challenging*, crossed that out, and kept writing, which seemed an appropriate summation of paleontology research as a whole. It is challenging. Never mind. Go on.

At some point, the screensaver on the computer started cycling through images of fossil slides. One looked like abstract art, with bright turquoise, red, and lavender stripes; another contained thick, orange-brown squiggles, like brushstrokes of characters in an unknown language; in another, violet, gold, and pale blue flecks shimmered through black splotches. The last image reminded me of something Kammerer had said: "We see the fossil record but through a glass darkly."

Over the next several months, the glass got even cloudier. It turned out that the preliminary results Meg had shown me were faulty because of inconsistencies in how she and the lab technician had counted stress lines. The early data analysis suggested that growth patterns were similar in Antarctic and South African tusks, but this finding didn't match what Meg saw by eye: thick bands in Antarctic fossils. It didn't make sense.

Meg ended up reanalyzing the tusks to count the lines more consistently. This time, she confirmed that Antarctic tusks, on average, had thicker and more frequent stress bands. But she didn't think this meant that *Lystrosaurus* was cold-blooded. The lines formed a distinctive pattern: multiple bands stacked close together. In searching the scientific literature, Meg realized that the stacked bands resembled those in the teeth of rodents, such as squirrels, during hibernation.

She felt "very happy," she said. "Because it just finally made sense to me." The similarities suggested that *Lystrosaurus* had gone through periods of torpor—that is, slowed down metabolism to conserve energy. The animals had likely hunkered down in their burrows, Meg said. The pattern of bands also indicated that *Lystrosaurus* periodically came out of torpor for brief growth spurts. These arousals are characteristic of hibernation in warm-blooded animals, she said; cold-blooded animals, in contrast, tend to completely "shut it down" during torpor. The data,

Meg concluded, supported the idea that *Lystrosaurus* was indeed warm-blooded.

Other scientists expressed caution, noting that the number of Antarctic tusks in the study was small, and that we needed more evidence from modern animals that cold-blooded creatures didn't arouse periodically as well. Even then, the question of whether *Lystrosaurus* was warm-blooded might not be definitively answered. Kammerer said, "I'm not sure it's knowable."

Though I didn't know this at the time, Ken Angielczyk and an international team of scientists had noticed a curious pattern in other therapsid fossils. They'd been examining parts of the animals' inner ears, called the semicircular canals. In dicynodonts, a group of animals that included *Lystrosaurus*, the canals were very oval-shaped; but in mammals, the canals tended to be more circular. Living reptiles also had oval canals. Why was that?

Angielczyk's team realized that there might be a connection between an animal's ear canal shape and body temperature. Hair cells detected the movement of fluid in the canals, which enabled animals to recognize their head movements and stay balanced. As a creature evolved from cold-blooded to warm-blooded, the fluid probably became runnier. For animals to track their movements properly, the physical properties of the fluid or the shape of the ear canal would need to change—and for mammals, it seemed to be the latter.

Based on canal shape patterns in modern animals, the researchers could deduce which extinct animals had likely been warm-blooded. In a paper published two years after Meg's study, they concluded that warm-bloodedness—the sustained high body temperatures seen in mammals today—had arisen about 233 million years ago, or 15 million years after *Lystrosaurus* had vanished.

I got back in touch with Meg, who was now running her own lab at Loyola University Chicago. She thought that the approach used in the ear canal study was new and unexpected and the evidence seemed sound. But as researchers try to figure out the origin of warm-bloodedness, "there are going to be multiple lines of evidence," she said. "There's not going to be a smoking gun." *Lystrosaurus* definitely had a high metabolic rate on par with warm-blooded animals today, she said, though exactly where it lay on the gradient from cold- to warm-blooded was up for debate. Meg still thought it was possible that *Lystrosaurus* had a high body temperature. "I like to keep the door open," she said.

Back in 2019, after one of my visits to Meg's lab, I had examined a photograph of a *Lystrosaurus* tusk on my computer. It showed one-quarter of the circular cross section, like a clock face from twelve to three. I could see an entire landscape within that fragment: brown layers resembling rough dirt cliffs, splotched with white, embedded with blue, yellow, and pink flecks, and splayed with cracks. It brought to mind a photo of Antarctica I'd seen in a book by another paleontologist who worked there in the late 1960s, which showed flat terrain, crisscrossed by trails, and low peaks in the background. Meg had looked for answers in the miniature landscape of the tusk tip, just as other scientists had searched those mountaintops for fossils fifty years earlier. The tusk was a world inside a world.

Then the world expanded again: the concentric circles of the fossil, I realized, looked like a cross section of our planet. I had a curious sensation of simultaneous enlargement and focus, space and time momentarily collapsing. It was easy to imagine that the lines were layers of stone and earth and minerals, and the pocket of gemlike colors in the center was the metal core. That the white patches near the surface were glaciers, and the jagged bumps were mountains, riddled with vanished creatures' bones. And that the tiny dark flecks clinging to the edge were people, Meg among them, digging and collecting and sorting and counting to try to get to the center of things.

Epilogue

An Act of Humility

Let me tell you a story.

In 1929, German meteorologist and geophysicist Alfred Wegener began preparing for an ambitious expedition to Greenland. He wanted to measure the thickness of the ice cap and collect atmospheric data, which would require working at sites on the island's coasts and a central station called Eismitte near the summit. But from the beginning, the fieldwork was plagued by delays: Thick sea ice prevented supplies from reaching his team on time, disorganized boxes made it difficult to find important items, motorized propeller sledges didn't perform well in fresh snow and could carry only about 1,100 pounds at a time. Worst of all, the two researchers who had traveled to Eismitte hadn't brought enough petroleum for heating and cooking to last the winter.

When Wegener, stationed on the coast, received word that his colleagues at the inland station needed more provisions, he set off with another scientist and a group of locals to deliver supplies. The cold was severe, and all but one of the Greenlanders, Rasmus Villumsen, turned back partway through the journey; by the time the remaining team reached Eismitte, the temperature had plunged below –50°F. Wegener and Villumsen began their return to the coast in November 1930, after celebrating Wegener's fiftieth birthday at the station with fruit and chocolate.

The two men never reached their destination. Wegener's body was later found tucked between sleeping bag covers, marked with a pair of skis presumably left by Villumsen, resting beneath a few feet of snow.

About two decades earlier, Wegener had made a radical proposal. He suggested that the continents, which had once been glommed together, broke into pieces that drifted long distances across the globe. This would explain why the edges of Africa and South America seem to fit together so nicely, a curious fact that scholars had remarked upon as far back as the late sixteenth century. In the mid-1700s, for instance, a German theologian wrote, "One supposes that perhaps both of these continents were previously attached to each other," and then separated when sea levels rose, or the legendary lost civilization of Atlantis sank. Wegener's theory would also account for scientists' discoveries of similar fossils and rock formations on different continents.

Other researchers had come up with different explanations for the fossil similarities. In the 1800s, scientists believed that Earth was cooling down and contracting. As the planet's core shrank, the thinking went, the surface crumpled. According to one theory, the lower folds became ocean basins, the upper ones turned into continents, and their positions periodically reversed as the wrinkling continued.

But new findings had started to poke holes in these ideas. In 1896, French physicist Henri Becquerel discovered radioactivity in uranium, a strange phenomenon that Marie and Pierre Curie investigated further. It turned out that the "decay" of radioactive elements generated heat. And since these elements were present in rocks, researchers began to question their understanding of the planet's cooling. "Are we living on a world heated throughout by radio-thermal actions?" wondered one geologist. "The inference that uranium exists yet in small quantities far down in the materials of the globe is highly probable... If it is present, then the central parts of the earth are rising in temperature." Meanwhile, other scientists found folds of rock in mountains that couldn't be explained by contraction alone.

Still, that didn't mean that geologists were ready to accept Wegener's idea that pieces of a supercontinent, later dubbed Pangaea, could drift apart "like pieces of a cracked ice floe in water," as he wrote. Though he had scattered supporters, many scientists in the United States dismissed him, calling his method "not scientific" and accusing him of taking "extraordinary liberties with the earth's rigid crust." In some circles, Wegener continued to be a laughingstock for decades. And yet evidence eventually emerged suggesting that continental drift might not be so absurd after all. From the 1930s to 1950s, scientists uncovered curious patterns in the ocean, ranging from magnetic "stripes" on the sea floor to a rift running down an underwater mountain range.

These findings could be explained by a new theory, which would also explain how continents drifted. Earth's surface was broken into gigantic plates that included both ocean floor and continental land. In the planet's interior, uranium and other radioactive elements generated extra heat. Plumes of hot, less dense rock rose to the surface, and lava burbled through the ocean floor along some plate edges. The lava piled up into peaks, started cooling, and sank as it moved outward to form long mountain ranges—essentially underwater volcanoes.

Meanwhile, in other parts of the oceans, the colder, denser edges of plates descended back into Earth's innards and gashed massive trenches. Scientists eventually figured out that this motion pulled the plates apart at their mountainous seams, where the rising lava continually filled in the gaps to become new sea floor.

In this way, an entire plate could slide along, with the cooler edge sinking into the planet's interior and the trailing edge being replenished by emerging hot rock. Any continental land that was part of the plate simply came along for the ride. Wegener was right: Continents could move. This change in scientific thinking was considered one of the biggest scientific revolutions of all time, on par with Darwin's theory of evolution.

Plate tectonics also helped scientists make sense of perplexing discoveries about animals. In the mid-1800s, the naturalist Alfred Russel Wallace traveled to the Malay Archipelago and noticed a stark difference in the creatures on the western and eastern islands. On Borneo, Sumatra, Java, and Bali, wild animals similar to those in mainland Asia roamed the forests: tapirs, monkeys, civets, woodpeckers. But on islands to the east, many animals resembled those in Australia, such as opossums, cockatoos, and brushturkeys.

"There is the greatest possible contrast in the animal productions," Wallace marveled, and the difference was as "zoologically wide as the poles asunder." He drew a boundary between the two sets of islands, which became known as the Wallace Line.

With plate tectonics, scientists could now envision how such different menageries had ended up right next to each other. In the distant past, the two groups of creatures had evolved on separate continents—a proto-Australia and proto-Asia—which were far apart on the globe. But over millions of years, the Australian land mass moved closer to the Asian one. The tectonic plates containing the two continents crashed into each other, triggering a chain of geological events that created new islands between them.

Meanwhile, other existing clusters of islands were connected to the Asian or Australian land masses during periods when sea levels were low, so they were filled with their continents' respective fauna. When Earth's glaciers melted and sea levels rose, water separated these pockets of land from the mainlands.

Animals flew, swam, or drifted between the continents and various islands. But many creatures' migrations across the Wallace Line might have failed when they encountered deep ocean, strong currents, unfavorable climate, or competition from existing species. What remained was what Wallace saw: some islands populated by Asian wildlife, next to other islands with abundant Australian fauna. Without plate tectonics, these creatures would have largely stayed put on their distant lands, never coming into close proximity.

By the mid-1960s, many scientists had accepted continental drift, but some skeptics remained. Then in 1967, a geologist discovered a jawbone fragment from an amphibian in Antarctica similar to others that had been excavated in southern Africa—a tantalizing piece of evidence that those two continents had been attached to each other in the past. Still, one could nitpick that the creature could have swum (with difficulty) across the ocean. So Edwin Colbert, a paleontologist at the American Museum of Natural History in New York City, and his team decided to pursue more convincing proof. In Antarctica, they would search for fossils of terrestrial animals found on other southern continents—creatures that definitely could not swim long distances.

After about a week of searching, Colbert thought, "Now, why can't we find a good bone of *Lystrosaurus*?" Many fossils of these creatures had been uncovered in Africa, and they would be relatively easy to identify. Shortly afterward, two team members dug fragments out of a sandstone ledge, including a chunk that appeared to be a tooth and part of a skull. When they brought their finds to Colbert, he saw that the fossil was clearly *Lystrosaurus*; within the bone, he later wrote, "was very definitely the end of a tusk!" The discovery was "indisputable" evidence that Antarctica had once been connected to Africa, he said. There was no way that this piglike creature could have paddled across the ocean.

After the rise of *Lystrosaurus* came the reign of dinosaurs, then mammals, and then humans. We won. We couldn't fly, but we invented planes. We couldn't swim long distances, but we invented ships. With our mechanical creations, we traversed the oceans with huge cargoes and blithely hopped around the world as we liked. We stitched the continents back together by sheer force of will, as if they were again one giant Pangaea.

With our superior cognition, we carved sculptures, designed skyscrapers, wrote concertos, devised medicines to cure deadly diseases, built telescopes so powerful that we could see signs of distant worlds hundreds of light-years away. We felled swaths of forest, built sprawling cities, cleared the way for more factories and farms. We domesticated billions of animals. We drove other species to extinction. We burned so much fossil fuel that we warmed the planet and lifted the seas, like gods. We harnessed the power of chemical elements extracted from underground, desecrating the land for generations to come. We poisoned ourselves. We ruled the earth.

As we claimed more habitat for our own and sold captured animals as wares, we brought humans and other creatures into closer quarters—creating the perfect conditions for viruses to spread from wildlife to people. The same planes and ships we had devised for our convenience carried pathogens around the globe just as easily. When SARS-CoV-2 emerged, it leaped with ease from country to country, from China to the United States and Uruguay and New Zealand and the Democratic Republic of the Congo, the oceans providing no barrier at all. While controversy swirled over whether the pathogen originated from a market or a lab, some biologists wondered: Did African swine fever, the virus that had wiped out wild pigs in Borneo, pave the way for the COVID-19 pandemic? Millions of pigs in China died from infection or were culled to stop the swine fever's spread, leading to a drastic drop in the pork supply and perhaps increasing the demand for wild animals, like those sold at the Huanan market in Wuhan. Whether the two pandemics were connected, we may never know.

At times, the story of our rise and fall took absurd turns. Wildfires were raging, storms pummeling coastlines, cities flooding. We wanted uranium to move away from fossil fuels to nuclear power; we wanted rare earth elements to build wind turbines and electric cars. The solution, some thought, was to restart uranium mining on the same desert land that we had yet to heal. In fact, with climate change melting the ice in the Arctic and opening shipping routes for more of the year, it was the perfect opportunity to propose opening a rare earth mine in Greenland, where uranium would also be extracted, next to an Indigenous community. We had learned nothing.

As for the future: Who knows? Perhaps another pandemic will wipe out most of humanity and the coyotes and raccoons will win—our cities merely shells for wild animals to wander, poking through the remains of our sandwich shops and banks and billionaires' mansions. Or perhaps our intelligence and endless creativity will conjure up the tools to bend the curve of rising temperatures back down and halt sea-level rise, like Poseidon raising his trident. In about 250 million years, some scientists predict, the continents will collide again and form a new supercontinent, Pangea Ultima: a hot, inhospitable land covered with thousands of miles of volcanoes, where temperatures in many areas could reach about 100°F to 120°F. Most mammals will likely die in a mass extinction. Perhaps nothing will be left except the preserved remains of our bones, our sculptures, our skyscrapers, and our telescopes for a visiting alien species from a distant planet to one day excavate and exclaim over—at what a flawed and clever species we were, capable of acts of such great destruction and intense beauty.

When I started writing this book, I wasn't aware of the overlaps between the scientists' stories. In retrospect, they should have come as no surprise: Of course the creatures in ecosystems are intertwined, our planet's geology drives the development of life, and the repercussions of humans' actions reverberate around the globe. It's a cliché to remark that everything is connected.

But seeing the researchers as part of a bigger story still gave me a new appreciation for their work. In isolation, it would be easy to condemn each project as too small, too insignificant to make a real difference. What matters is that these scientists keep pursuing answers, that they throw themselves into the breach over and over, in spite of the fact that they will likely play very minor roles—that they'll each carve out and chip away at their own little fragment of colored glass, turning it and peering at it from all angles, and then finally fit their minuscule tile into a larger mosaic where it will, from a distance, become nearly invisible. Doing science is an act of humility. It's also an act of hope, that everyone's collective work will still matter in the end.

From my vantage point, I can't say whether the research I've described in these chapters will be important. The reward for surviving the messy middle of science usually isn't Truth with a capital T or a Nobel Prize–worthy discovery; it's a tentative step forward, made with the knowledge that someone else can always pry out that exquisitely-crafted tile, examine it under a magnifying glass, pronounce it unconvincing, and toss it aside. I don't know which findings will survive scrutiny or how much they'll shape our understanding of our world. Will they matter? Perhaps.

But even if they don't, the scientists' deeply human attempts—the curiosity, the questioning, the desire to change our trajectory, the *trying*—do, I think.

I hope that they will always keep trying.

Acknowledgments

My formidable agent, Alice Martell, believed in this book from day one. Her energy and enthusiasm lit a fire under this project, and Stephanie Finman and Teresa Cavanaugh at The Martell Agency provided much-appreciated support behind the scenes.

Many thanks to my editor Jenna Jankowski for her insightful comments, eternally cheerful attitude, and uncanny ability to see and crystallize the book from a thirty-thousand-foot view, as well as for juggling the many balls needed to get the manuscript over the finish line. Erin McClary first saw the potential in my proposal and offered thoughtful edits on early chapter drafts. Turning a humble Word document into a Real Book is an act of magic, and I'm immensely grateful to the team at Sourcebooks that made it happen, including Emily Proano, Ariel Curry, Mary Wheelehan, Jillian Rahn, Nia Saxon, Kirsten Clawson, Tara Jaggers, and Kelly Lawler. Copy editor Lisa Theobald saved me from countless grammatical errors, and proofreader Sharon Sofinski buffed the text to a high shine. Thank you also to Angela Corpus, Jennifer Steinhagen, and everyone else on the marketing team for the critical work they do to connect authors with readers and keep physical books alive.

Collaborating with artist Emma Regnier was a dream come true; her elegant, heartfelt illustrations gave the book its soul. Caitlin Sacks designed a stunning cover that perfectly captured the themes of wonder, curiosity, and discovery.

This book would not exist without the generous funding I received from the Knight Science Journalism Program at the Massachusetts Institute of Technology (MIT) and the Alfred P. Sloan Foundation. Rob Irion set the entire project into motion when he kindly wrote me a reference letter for the MIT fellowship, and Deborah Blum and Ashley Smart offered wise and gentle guidance during that cherished year. Courtney Maum, the "Query Doula," gave astute feedback on the book proposal that emerged.

Fact-checkers Tina Knezevic, Emily Krieger, and Lindsay Gellman took a tremendous load off my shoulders with their razor-sharp attention to detail, good humor, and willingness to dive into everything from glacial hydrology to soft matter physics. I will never mix up a Mets and Yankees baseball cap again. Any remaining errors are, of course, my own.

Janina Lawrence, Joseph Lee, Tyler Santora, and María Paula Rubiano A. provided keen and exacting sensitivity reading of several chapters. Thank you also to María Paula and Peter Kum for their help with Spanish and French translation, respectively.

Even though AI is now taking over everything, it doesn't hold a candle to the expert and accurate transcription by Joanna Parson and Nico Ager at Letter Perfect Transcription and by Courtney Columbus. Their transcripts of long, complex, highly technical interviews saved my sanity.

Thank you to Ada Limón for allowing me to reprint the lines

from her wonderful poem and to Navajo Nation TV & Film for giving me permission to report in Cameron and Tuba City.

I would be nowhere without my fellow writers. Sarah DeWeerdt, Eric Wagner, Deirdre Lockwood, Michael Bradbury, Ashley Braun, Wayt Gibbs, Sally James, Ellen Kuwana, Michelle Martin, Bryn Nelson, Bonnie Rochman, Paula Carter, Chris Harvey Bate, Megan Savage, and Jane Liaw generously offered support and feedback on my proposal, drafts, and half-baked ideas at various points in the process. Amber Dance provided incisive developmental editing on an early chapter and listened to me drone on about endless obstacles during our weekly calls. Lynne Peeples patiently helped me navigate unfamiliar debut-author terrain. Christie Aschwanden's guidance and support for other freelancers is unparalleled; I also got a torrent of good advice from the Gentlewomen, Globalance, and the book-writing Slack channel.

My journey toward nonfiction writing and science journalism started with Scott Russell Sanders and Holly Stocking, who opened my eyes to a new way of seeing and writing about the world. The year that I spent at the University of California, Santa Cruz's science communication program was one of the most grueling, enlightening, and transformative of my life. The stellar editors at *Nature*, including Alexandra Witze, Rich Monastersky, and Brendan Maher, encouraged and honed my feature writing; Karen Kaplan (1959–2023), the indefatigable Careers section editor, believed in the importance of junior scientists' stories and was a staunch champion for her writers.

I interviewed so many researchers who ultimately were not quoted in the book, and I am incredibly grateful to all of them for their time and expertise. Particular thanks go to Hannah

Sophia Davies, Brad Foley, Christopher Spencer, Mark Behn, and Jon Powell, who endured my confusion and unending questions about plate tectonics; ditto for Larry Heaney on the Wallace Line.

Finally, thank you to my parents, Emily and Sun, and my sister, Kelly, for their support; to my husband, Chris, for holding down the fort during my reporting travels; and to my daughter, Gillian, who once paid me the highest compliment a science writer can receive: "You explain things good."

Notes

Direct and paraphrased quotes were drawn from interviews conducted over the phone or Zoom, in-person reporting, or email communication, unless otherwise indicated in the endnotes. Similarly, details and dialogue in narrative scenes were drawn from in-person reporting trips or from sitting in on team Zoom meetings. Retrospective narrative scenes (where I was not present) were reconstructed through interviews with the team members involved.

For dialogue drawn from in-person reporting and Zoom meetings, I sometimes rearranged the chronology of conversations to make them easier for readers to understand. For example, if scientists discussed topics A, B, and C, then returned to talking about A, then C, then back to A again, I might describe all the topic A material in one paragraph, then B in the next paragraph, and then C (provided that this reordering did not change the meaning of what they said due to differences in context).

These citations are not a comprehensive list of all the sources I consulted. In some cases, I drew on several sources to write a particular sentence and listed one or two of the most relevant.

Introduction: Into the Wardrobe

Details about the sunflower research were drawn from interviews with Ben Blackman, my in-person visits to the Indiana University Bloomington lab, or the scientific paper Benjamin K. Blackman et al., "Contributions of Flowering Time Genes to Sunflower Domestication and Improvement," *Genetics* 187, no. 1 (January 1, 2011): 271–87, https://doi.org/10.1534/genetics.110.121327. Details about other research projects were drawn from my reporting trips to those locations, interviews with the scientists involved, or Zoom meetings that I attended, all explained and cited in greater detail in upcoming chapters.

The word "science" often brings: Joanna Huxster (researcher who studies public understanding of ideologically divisive science, Eckerd College), in discussion with the author, July 6, 2023; Anna Marie Roos (historian, University of Lincoln), email message to author, July 12, 2023.

Researchers spend most of their time: For example, see Michael Strevens, *The Knowledge Machine: How Irrationality Created Modern Science* (Liveright, 2020): 33–35.

"Eureka!" means "I have": *Oxford English Dictionary*, s.v. "Eureka," accessed April 25, 2025, https://www.oed.com/search/dictionary/?scope=Entries&q=eureka; David Biello, "Fact or Fiction?: Archimedes Coined the Term 'Eureka!' in the Bath," *Scientific American*, December 8, 2006, https://www.scientificamerican.com/article/fact-or-fiction-archimede.

Frodo and Sam had delivered: J. R. R. Tolkien, *The Return of the King* (1955; repr., Unwin Paperbacks, 1987), 7, 252, 270–71; C. S. Lewis, *The Complete Chronicles of Narnia* (1950; repr., HarperCollins, 1998): 77, 86–87, 100–105, 115, 130.

more than 600 light-years: "WASP-49 b," NASA, accessed April 25, 2025, https://science.nasa.gov/exoplanet-catalog/wasp-49-b.

the Warren-girder case: Madelyn Jane Leembruggen, "Buckling, Wrinkling, and Crumpling of Simulated Thin Sheets" (PhD diss., Harvard University, 2024), 120; Madelyn Leembruggen (physicist and science communicator), in discussion with the author, September 21, 2021.

To uncover part of: Megan R. Whitney and Christian A. Sidor, "Evidence of Torpor in the Tusks of *Lystrosaurus* from the Early Triassic of Antarctica," *Communications Biology* 3, no. 1 (August 27, 2020): 471, https://doi.org/10.1038/s42003-020-01207-6; Meg Whitney (vertebrate paleontologist, Loyola University Chicago), in discussion with the author, November 28, 2018; Gage Moreno (viral genomicist, Broad Institute), in discussion with the author, March 22, 2022; Katarina Braun (resident in obstetrics, gynecology, and reproductive sciences, Yale New Haven Health), in discussion with the author, March 31, 2022.

As philosopher of science: Strevens, *The Knowledge Machine*, 12.

what poet Stanley Plumly: Maggie Smith, *You Could Make This Place Beautiful: A Memoir* (Atria/One Signal, 2023), 306; "Stanley Plumly," Poetry Foundation, accessed April 25, 2025, https://www.poetryfoundation.org/poets/stanley-plumly.

"fixers" and "dreamers": We can quibble over whose work is "useful" or worthy, but I won't do that here. I will, however, point you to David Samuel Shiffman, "Why Are We Funding This?," *American Scientist* 113, no. 4 (July–August 2025): 220, https://www.americanscientist.org/article/%E2%80%9Cwhy-are-we-funding-this%E2%80%9D. Shiffman explains how seemingly "silly"-sounding science can lead to unexpected societal benefits—for instance, the study of lizard venom laying the groundwork for weight-loss drugs or bee foraging research inspiring an algorithm to manage internet traffic—and the ways that research universities drive local economies, particularly in rural areas.

another researcher told me: Christian Kammerer (paleontologist, North Carolina Museum of Natural Sciences), in discussion with the author, September 12, 2019; *The Holy Bible: King James Version* (1611; repr., Hendrickson Publishers, 2004), 553, https://www.google.com/books/edition/The_Holy_Bible/3EUgOyrQnmEC.

Geology: The Invisible River

Details about the team's Helheim Glacier research and fieldwork were drawn from interviews with team members Jessica Mejia, Kristin Poinar, Courtney Shafer, Winnie Chu, Renée Clavette, TJ Young, Angelo Tarzona, Colin Meyer, Aleah Sommers, Ian McDowell, Ilyse Horlings, Mike Coyle, and Richard Mansfield; emails, text messages, voice memos, photos, and videos from team members; interviews with helicopter pilots Jean-Marie Bärtsch and Samuel Müller; my reporting trip to Greenland; team Zoom meetings that I attended; research and fieldwork planning documents that the team shared with me; and the scientific paper Jessica Z. Mejia et al., "Mechanisms for Upstream Migration of Firn Aquifer Drainage: Preliminary Observations from Helheim Glacier, Greenland," *Journal of Glaciology* 71 (2025): e5, https://doi.org/10.1017/jog.2024.78. Biographical details about Mejia, Poinar, and Chu were drawn from interviews with them. Information about ice sheets, sea-level rise, and firn aquifers was drawn from interviews with team members or experts such as Michiel van den Broeke, Amber Leeson, Mike MacFerrin, Ginny Catania, Leigh Stearns, Mathieu Morlighem, Clément Miège, Rick Forster, Lynn Montgomery, Olivia Miller, Celia Trunz, and Joanna Huxster.

named for a Norse world: Theresa Bane, *Encyclopedia of Mythological Objects* (McFarland, 2020), 86, https://www.google.com/books/edition/Encyclopedia

_of_Mythological_Objects/RAnoDwAAQBAJ; R. B. Anderson, *Norse Mythology; Or, the Religion of Our Forefathers*, 2nd ed. (Chicago, 1876; Project Gutenberg, 2021), chap. 7, https://www.gutenberg.org/files/65910/65910-h/65910-h.htm.

about sixty-five miles inland: Jean-Marie Bärtsch (helicopter pilot, GreenlandCopter), WhatsApp message to author, April 29, 2025.

Helheim Glacier was an unforgiving: Information about Helheim was synthesized from Aleah Sommers et al., "Subglacial Hydrology Modeling Predicts High Winter Water Pressure and Spatially Variable Transmissivity at Helheim Glacier, Greenland," *Journal of Glaciology*, June 21, 2023, 1–13, https://doi.org/10.1017/jog.2023.39; M. Morlighem et al., "BedMachine v3: Complete Bed Topography and Ocean Bathymetry Mapping of Greenland from Multibeam Echo Sounding Combined with Mass Conservation," *Geophysical Research Letters* 44, no. 21 (November 16, 2017), https://doi.org/10.1002/2017GL074954; Mathieu Morlighem (glaciologist, Dartmouth College), email message to author, January 30, 2025; Aleah Sommers (postdoctoral researcher, Dartmouth College), in discussion with the author, September 27, 2022, and August 18, 2023; Courtney Shafer (PhD student, University at Buffalo), in discussion with the author, September 26, 2022; Jessica Mejia (postdoctoral researcher, University at Buffalo), in discussion with the author, May 10, 2023; Colin Meyer (fluid dynamicist, Dartmouth College), in discussion with the author, September 21, 2022.

In 2011, researchers had discovered: The story of the firn aquifer discovery was synthesized from Richard R. Forster et al., "Extensive Liquid Meltwater Storage in Firn within the Greenland Ice Sheet," *Nature Geoscience* 7 (2014): 95–98, https://doi.org/10.1038/ngeo2043; Rick Forster (glaciologist, University of Utah), in discussion with the author, May 30, 2023; Clément Miège (air monitoring specialist, Puget Sound Clean Air Agency), in discussion with the author, March 31, 2023, and November 25, 2024; Evan Burgess (lead data scientist, Airborne Snow Observatories), email message to author, January 30, 2025; Terrance Gacke (former field engineer, University of Alaska Fairbanks), email messages to author, January 30 to February 3, 2025.

Burgess and Miège had uncovered: Forster et al., "Extensive Liquid Meltwater Storage"; Joel Harper, "Greenland's Lurking Aquifer," *Nature Geoscience* 7 (2014): 86–87, https://doi.org/10.1038/ngeo2061.

Later measurements suggested: Lynn N. Montgomery et al., "Investigation of Firn Aquifer Structure in Southeastern Greenland Using Active Source Seismology," *Frontiers in Earth Science* 5 (February 7, 2017), https://doi.org/10.3389/feart.2017.00010; Miège, email message to author, January 30, 2025; Miège, discussion, November 25, 2024.

Radar data revealed signs: Clément Miège et al., "Spatial Extent and Temporal Variability of Greenland Firn Aquifers Detected by Ground and Airborne Radars," *Journal of Geophysical Research: Earth Surface* 121, no. 12 (December 2016): 2381–98, https://doi.org/10.1002/2016JF003869; Miège, discussion, March 31, 2023, and November 25, 2024; Olivia Miller (former PhD student, University of Utah), in discussion with the author, March 22, 2023; Lynn Montgomery (senior research scientist, Lockheed Martin), in discussion with the author, April 11, 2023.

The water was stored in firn: Forster et al., "Extensive Liquid Meltwater Storage"; Mejia, discussion, January 15, 2021; Winnie Chu (glacial geophysicist, Georgia Tech), in discussion with the author, August 4, 2021; Amber Leeson (glaciologist, Lancaster University), in discussion with the author, March 20, 2023; Kristin Poinar (glaciologist, University at Buffalo), in discussion with the author, September 22, 2021.

The discovery raised questions: Miller, discussion; Chu, discussion; Miège, discussion, March 31, 2023; Lora S. Koenig et al., "Initial In Situ Measurements of Perennial Meltwater Storage in the Greenland Firn Aquifer," *Geophysical Research Letters* 41 (2014): 81–85, doi:10.1002/2013GL058083.

Kristin Poinar, a glaciologist: Poinar, discussion; Kristin Poinar et al., "Drainage of Southeast Greenland Firn Aquifer Water through Crevasses to the Bed," *Frontiers in Earth Science* 5, article 5 (2017), doi: 10.3389/feart.2017.00005.

predictions of sea-level rise: Details about uncertainty in sea-level rise projections were synthesized from H. Lee and J. Romero, eds., "Climate Change 2023: Synthesis Report. Contribution of Working Groups I, II and III to the Sixth Assessment Report of the Intergovernmental Panel on Climate Change" (IPCC, 2023), 35–115, doi:10.59327/IPCC/AR6-9789291691647; H. Lee and J. Romero, eds., "Summary for Policymakers. Contribution of Working Groups I, II and III to the Sixth Assessment Report of the Intergovernmental Panel on Climate Change" (IPCC, 2023), 1–34, doi:10.59327/IPCC/AR6-9789291691647.001; Andy Aschwanden et al., "Brief Communication: A Roadmap Towards Credible Projections of Ice Sheet Contribution to Sea Level," *The Cryosphere* 15, no. 12 (December 17, 2021): 5705–15, https://doi.org/10.5194/tc-15-5705-2021; Michiel van den Broeke (polar climatologist, Utrecht University), in discussion with the author, April 19, 2023, and November 27, 2024; Mike MacFerrin (glaciologist, University of Colorado Boulder), in discussion with the author, March 17, 2023; Chu, discussion; Leeson, discussion; Leigh Stearns (glaciologist, University of Kansas), in discussion with the author, April 13, 2023; Forster, discussion; Mejia, discussion, January 15, 2021.

The effect of Greenland's firn: Koenig et al., "Initial In Situ Measurements";

van den Broeke, discussion, April 19, 2023, and November 27, 2024; Stearns, discussion.

If the entire Greenland ice: Morlighem et al., "BedMachine v3"; Michael Oppenheimer et al., "Sea Level Rise and Implications for Low-Lying Islands, Coasts and Communities," in *IPCC Special Report on the Ocean and Cryosphere in a Changing Climate*, eds. H-O. Pörtner et al. (Cambridge University Press, 2019), 330–32; van den Broeke, discussion, November 27, 2024.

In 2018, researchers confirmed: Lynn Montgomery et al., "Hydrologic Properties of a Highly Permeable Firn Aquifer in the Wilkins Ice Shelf, Antarctica," *Geophysical Research Letters* 47, no. 22 (November 28, 2020), https://doi.org/10.1029/2020GL089552; Montgomery, discussion; Miège, discussion, March 31, 2023; van den Broeke, discussion, April 19, 2023.

Firn aquifers form when: van den Broeke, discussion, April 19, 2023, and November 27, 2024.

If the water in firn aquifers: Leeson, discussion.

disintegration of one critical: Sainan Sun et al., "Antarctic Ice Sheet Response to Sudden and Sustained Ice-Shelf Collapse (ABUMIP)," *Journal of Glaciology* 66, no. 260 (December 2020): 891–904, https://doi.org/10.1017/jog.2020.67; Leeson, email message to author, April 7, 2025.

A mandala was an elaborate: "Hayagriva Mandala," Denver Art Museum, accessed April 21, 2025, https://www.denverartmuseum.org/en/edu/object/hayagriva-sand-mandala; "Tibetan Healing Mandala," Smithsonian Institution, accessed April 21, 2025, https://archive.asia.si.edu/exhibitions/online/mandala/faq.htm; "Learn More: The Tradition of the Sand Mandala," Ithaca College, accessed April 21, 2025, https://www.ithaca.edu/tibetan-buddhist-mandala-project/learn-more-tradition-sand-mandala; Carly Swain, "Tibetan Buddhist Monk Creates Sand Mandala as Exercise in Enlightenment," University of North Carolina at Chapel Hill, February 28, 2017, https://global.unc.edu/news-story/tibetan-buddhist-monk-creates-sand-mandala-as-exercise-in-enlightenment-2.

town of about eighteen hundred people: "Population in Localities January 1st 1977–2025," Statbank Greenland, accessed April 21, 2025, https://bank.stat.gl/pxweb/en/Greenland/Greenland__BE__BE01__BE0120/BEXSTD.px; "Tasiilaq," Trap Greenland, accessed April 21, 2025, https://trap.gl/en/kommunerne-og-byerne/kommuneqarfik-sermersooq/tasiilaq.

Ice sheets might look: Details about ice sheet movement were synthesized from Leeson, discussion; Mejia, discussion, January 15, 2021, and January 11, 2022; Renée Clavette (graduate student, Georgia Tech), in discussion with the author, April 13, 2022; van den Broeke, discussion, November 27, 2024; Poinar, discussion.

For sea-level rise projections: The two paragraphs about ice sheet mass loss and scientists' perspectives in the 1990s were synthesized from van den Broeke, discussion, April 19, 2023, and November 27, 2024; Leeson, discussion; Mejia, discussion, January 15, 2021, and March 28, 2022; MacFerrin, discussion; *Climate Change: The 1990 and 1992 IPCC Assessments* (Intergovernmental Panel on Climate Change, 1992): 52–53, 81, 110.

in the mid-2000s, satellite data: Isabella Velicogna and John Wahr, "Greenland Mass Balance from GRACE," *Geophysical Research Letters* 32, no. 18 (September 28, 2005): 2005GL023955, https://doi.org/10.1029/2005GL023955; Isabella Velicogna and John Wahr, "Measurements of Time-Variable Gravity Show Mass Loss in Antarctica," *Science* 311, no. 5768 (March 24, 2006): 1754–56, https://doi.org/10.1126/science.1123785.

In Antarctica, warmer ocean water: MacFerrin, discussion; van den Broeke, discussion, November 27, 2024.

over the next couple of decades: Leeson, discussion; MacFerrin, discussion; Meyer, discussion, June 29, 2022.

During the summer of 2012: S. V. Nghiem et al., "The Extreme Melt across the Greenland Ice Sheet in 2012," *Geophysical Research Letters* 39, no. 20 (October 28, 2012): 2012GL053611, https://doi.org/10.1029/2012GL053611; Johanna Beckmann and Ricarda Winkelmann, "Effects of Extreme Melt Events on Ice Flow and Sea Level Rise of the Greenland Ice Sheet," *The Cryosphere* 17, no. 7 (July 27, 2023): 3083–99, https://doi.org/10.5194/tc-17-3083-2023.

Seven years later: Beckmann and Winkelmann, "Effects of Extreme Melt Events"; Von Walden et al., "Mechanisms of Multiple, Anomalous Melt Events at Summit Station, Greenland in Summer 2019," EGU General Assembly 2021 (April 2021), https://doi.org/10.5194/egusphere-egu21–13119.

That meltwater could have: The three paragraphs about meltwater effects were synthesized from H. Jay Zwally et al., "Surface Melt-Induced Acceleration of Greenland Ice-Sheet Flow," *Science* 297 (2002): 6; Ginny Catania (glaciologist, University of Texas at Austin), in discussion with the author, March 30, 2023; TJ Young (postdoctoral researcher, University of Cambridge), in discussion with the author, April 19, 2023; Chu, discussion; Mejia, discussion, January 15, 2021; Sommers, discussion, July 14, 2022; MacFerrin, discussion; Meyer, discussion, June 29, 2022; Poinar, discussion, September 22, 2021.

Jess would later tell me: Mejia, discussion, October 5, 2023.

Most of Tasiilaq's residents: Ida Moltke et al., "Uncovering the Genetic History of the Present-Day Greenlandic Population," *The American Journal of Human Genetics* 96, no. 1 (January 2015): 54–69, https://doi.org/10.1016/j.ajhg.2014.11.012; Susanne Houd et al., "Giving Birth in Rural Arctic *Greenland* Results from an Eastern Greenlandic Birth Cohort," *International Journal of Circumpolar Health* 81, no. 1 (December 31, 2022): 2091214, https://doi.org/10.1080/22423982.2022.2091214; Andrew J. Dugmore et al., "Cultural Adaptation, Compounding Vulnerabilities and Conjunctures in Norse Greenland," *Proceedings of the National Academy of Sciences* 109, no. 10 (March 6, 2012): 3658–63, https://doi.org/10.1073/pnas.1115292109; Mejia, discussion, October 5, 2023.

But people had started: Mejia, discussion, October 5, 2023.

Captain Kirk, who famously: "Just How Heroic is *Star Trek*'s 'I Don't Like to Lose' James T. Kirk?," *PopMatters*, October 7, 2016, https://www.popmatters.com/just-how-heroic-is-star-treks-i-dont-like-to-lose-james-t-kirk-2495413884.html.

When I'd talked to other: Leeson, discussion; Catania, discussion; Stearns, discussion.

Once they reached the ice layer: Ian McDowell (PhD student, University of Nevada), in discussion with the author, June 23, 2023.

At the bottom of the glacier: J. Z. Mejia et al., "Isolated Cavities Dominate Greenland Ice Sheet Dynamic Response to Lake Drainage," *Geophysical Research Letters* 48, no. 19 (October 16, 2021), https://doi.org/10.1029/2021GL094762; Sommers et al., "Subglacial Hydrology Modeling"; Sommers, discussion, August 18, 2023; Mejia, discussion, July 26, 2023.

Evolutionary Biology: Call of the Guardian Frog

Details about the smooth guardian frog research were drawn from interviews, email, or text correspondence with Johana Goyes Vallejos, Kentwood Wells, Ulmar Grafe, Hanyrol Ahmad Sah, or Joremy anak Tony; Goyes Vallejos' photos or videos; or from the scientific papers Johana Goyes Vallejos et al., "Calling Behavior of Males and Females of a Bornean Frog with Male Parental Care and Possible Sex-Role Reversal," *Behavioral Ecology and Sociobiology* 71, no. 6 (June 2017): 95, https://doi.org/10.1007/s00265-017-2323-3; Johana Goyes Vallejos, T. Ulmar Grafe, and Kentwood D. Wells, "Prolonged Parental Behaviour by Males of *Limnonectes Palavanensis* (Boulenger 1894), a Frog with Possible Sex-Role Reversal," *Journal of Natural History* 52, no. 37–38 (October 10, 2018): 2473–85, https://doi.org/10.1080/00222933.2018.1539196; and Johana Goyes Vallejos, T. Ulmar Grafe, and Kentwood D. Wells, "Factors Influencing Tadpole Deposition Site Choice in a Frog with Male Parental Care: An Experimental Field Study," *Ethology* 125, no. 1 (January 2019): 29–39, https://doi.org/10.1111/eth.12820. Biographical details about Goyes Vallejos were drawn from interviews with her and from her first-person story at "Human Nature: Coming of Age Stories," *The Story Collider*, July 9, 2021, https://www.storycollider.org/stories/2021/7/6/human-nature-coming-of-age-stories. General background about frog parental care and sex-role reversal was drawn from interviews with team members or experts such as Sarah Bush, Eva Fischer, Eva Ringler, Lisa Schulte, Kyle Summers, James Tumulty, Adam Jones, Jonathan Henshaw, Robin Hare, Malin Ah-King, and Sara Lipshutz. Information about Borneo was drawn from interviews with Grafe, Chien Lee, Jennifer Sheridan, and Poline Bala.

In 1984, a young man: The local Iban worker was not credited in the team's report, and I wasn't able to find out his name. However, in another paper summarizing multiple expeditions to Borneo, the researchers wrote the following: "We wish to express our thanks first to men of the Iban longhouse, Rumah Jimbong. Without their able assistance, none of the work reported here would have taken place" (Robert F. Inger, Harold K. Voris, and Karl J. Frogner, "Organization of a Community of Tadpoles in Rain Forest Streams in Borneo," *Journal of Tropical Ecology* 2, no. 3 [1986]: 204). The description of the 1984 expedition and discovery was drawn from several additional sources, including Robert F. Inger, Harold K. Voris, and Paul Walker, "Larval Transport in a Bornean Ranid Frog," *Copeia* 1986, no. 2 (May 9, 1986): 523, https://doi.org/10.2307/1445012; Harold K. Voris and Robert F. Inger, "Frog Abundance along Streams in Bornean Forests," *Conservation Biology* 9, no. 3 (June 1995): 679–83, https://doi.org/10.1046/j.1523–1739.1995.09030679.x; and Harold Voris (curator emeritus, Field Museum), in discussion with the author, October 31, 2024. Description of the smooth guardian frog was drawn from Goyes Vallejos, Grafe, and Wells, "Prolonged Parental Behavior"; Kentwood Wells (herpetologist, University of Connecticut), in discussion with the author, 2018; and photographs provided by Voris. The nineteenth-century report is G. A. Boulenger, "XIV.—On the Herpetological Fauna of Palawan and Balabac," *Annals and Magazine of Natural History* 14, no. 80 (August 1894): 81–89, https://doi.org/10.1080/00222939408677772.

Two years later: Description of the 1986 expedition was synthesized from Robert F. Inger and Harold K. Voris, "Taxonomic Status and Reproductive Biology of Bornean Tadpole-Carrying Frogs," *Copeia* 1988, no. 4 (December 28, 1988): 1060, https://doi.org/10.2307/1445733; Voris and Inger, "Frog Abundance"; Sharon B. Emerson (evolutionary biologist, retired), in discussion with the author, November 5, 2024; Voris, discussion. Emerson was one of the scientists involved in the survey.

They wrote up: Inger, Voris, and Walker, "Larval Transport"; Inger and Voris, "Taxonomic Status."

for the next few decades: Johana Goyes Vallejos (evolutionary behavioral ecologist, University of Missouri), in discussion with the author, March 30, 2023; Wells, discussion.

He found it intriguing: Recently, scientists have determined that smooth guardian frogs, previously called *Limnonectes palavanensis*, are actually a set of several closely related species, not just one species. However, all of these "sibling" species likely share some behaviors. Ulmar Grafe (behavioral ecologist, Universiti Brunei Darussalam), in discussion with the author, January 30, 2024; J. Maximilian Dehling et al., "Cryptic Radiation within the Tadpole-Carrying Guardian Frogs from Borneo, *Limnonectes palavanensis* and *L. finchi* (Anura: Dicroglossidae), with the Description of Eight New Species," *Zootaxa* 5650, no. 1 (2025): 1–80, https://doi.org/10.11646/zootaxa.5650.1.1.

Many scientists had observed: Kyle Summers and James Tumulty, "Parental Care, Sexual Selection, and Mating Systems in Neotropical Poison Frogs," in *Sexual Selection* (Elsevier, 2014), 289–320, https://doi.org/10.1016/B978-0-12-416028-6.00011-6; James Tumulty (behavioral ecologist, Rhodes College), in discussion with the author, February 16, 2024, and October 30, 2024.

Unlike the brightly colored: Summers and Tumulty, "Parental Care"; Kyle Summers (evolutionary biologist, East Carolina University), in discussion with the author, November 27, 2018.

these creatures are vulnerable, shy, and secretive: Goyes Vallejos, in discussion with the author, November 20, 2018, and March 30, 2023; Grafe, discussion; Wells, discussion.

This was unusual: The information about female frog calls that had been reported as of 2011, when Goyes Vallejos was beginning her project, was synthesized from Doris Preininger et al., "Comparison of Female and Male Vocalisation and Larynx Morphology in the Size Dimorphic Foot-Flagging Frog Species *Staurois Guttatus*," *Herpetological Journal* 26, no. 3 (July 2016): 187–97; Luís F. Toledo et al., "The Anuran Calling Repertoire in the Light of Social Context," *Acta Ethologica* 18, no. 2 (June 2015): 87–99, https://doi.org/10.1007/s10211-014-0194-4; Bert Willaert et al., "A Unique Mating Strategy Without Physical Contact During Fertilization in Bombay Night Frogs (*Nyctibatrachus Humayuni*) with the Description of a New Form of Amplexus and Female Call," *PeerJ* 4 (June 14, 2016): e2117, https://doi.org/10.7717/peerj.2117; Sharon B. Emerson and Sunny K. Boyd, "Mating Vocalizations of Female Frogs: Control and Evolutionary Mechanisms," *Brain, Behavior and Evolution* 53, no. 4 (1999): 187–97, https://doi.org/10.1159/000006594; and Kentwood D. Wells, *The Ecology and Behavior of Amphibians* (University of Chicago Press, 2007), 282–88. Since Goyes Vallejos began her project in 2011, the number of reported frog species in which females call as part of the mating process has risen to nearly forty; see Erika M. Santana, Angela M. Mendoza-Henao, and Johana Goyes Vallejos, "The 'Silent' Half: Diversity, Function and the Critical Knowledge Gap on Female Frog Vocalizations,"

Proceedings of the Royal Society B: Biological Sciences 292, no. 2047 (May 2025): 20250454, https://doi.org/10.1098/rspb.2025.0454. The number of frog and toad species at the time was provided by Darrel Frost (curator emeritus, American Museum of Natural History), email message to author, October 25, 2024; "Anura," Amphibian Species of the World, accessed October 25, 2024, https://amphibiansoftheworld.amnh.org/Amphibia/Anura.

a rare case of a phenomenon: The information about sex-role reversal in this paragraph was synthesized from Karoline Fritzsche et al., "The 150th Anniversary of the Descent of Man: Darwin and the Impact of Sex-Role Reversal on Sexual Selection Research," *Biological Journal of the Linnean Society* 134, no. 3 (October 22, 2021): 525–40, https://doi.org/10.1093/biolinnean/blab091; Jonathan Henshaw (theoretical evolutionary biologist, University of Freiburg), in discussion with the author, February 1, 2024; Robin Hare (lecturer, University of Western Australia), in discussion with the author, February 15, 2024; Adam Jones (evolutionary biologist, University of Idaho), in discussion with the author, November 28, 2018.

she'd published new findings: Goyes Vallejos, Grafe, and Wells, "Prolonged Parental Behaviour."

Frogs have a reputation: Eva Ringler et al., "What Amphibians Can Teach Us About the Evolution of Parental Care," *Annual Review of Ecology, Evolution, and Systematics* 54, no. 1 (November 2, 2023): 43–62, https://doi.org/10.1146/annurev-ecolsys-102221–050519.

around 10 to 25 percent: Ringler et al., "What Amphibians Can Teach"; Isabella Capellini, email message to author, February 28, 2025; Martha L. Crump, "Anuran Reproductive Modes: Evolving Perspectives," *Journal of Herpetology* 49, no. 1 (March 2015): 1–16, https://doi.org/10.1670/14–097; Wells, *Ecology and Behavior*, 517.

Scientists started noticing: The information about Merian and her observations was drawn from Maria Sibylla Merian, *Metamorphosis Insectorum Surinamensium: Ofte Verandering der Surinaamsche Insecten* (Amsterdam: 1705), 59; Londa Schiebinger, "Exotic Abortifacients and Lost Knowledge," *Lancet* 187, no. 6 (2008): 1350–53, https://doi.org/10.1016/j.cell.2024.01.045; Lisa M. Schulte et al., "Developments in Amphibian Parental Care Research: History, Present Advances, and Future Perspectives," *Herpetological Monographs* 34, no. 1 (July 15, 2020): 71–97, https://doi.org/10.1655/HERPMONOGRAPHS-D-19–00002.1; "Metamorphosis Insectorum Surinamensium," Royal Collection Trust, accessed March 12, 2025, https://www.rct.uk/collection/1085787/metamorphosis-insectorum-surinamensium; Lisa Schulte (biologist, Goethe University Frankfurt), in discussion with the author, February 15, 2024. The horror film quote is from Eva K. Fischer, "Form, Function, Foam: Evolutionary Ecology of Anuran Nests and

Nesting Behaviour," *Philosophical Transactions of the Royal Society B: Biological Sciences* 378, no. 1884 (August 28, 2023): 8, https://doi.org/10.1098/rstb.2022.0141.

In the late 1700s: Demours, P., "Observation au Sujet de Deux Animaux, Dont Le Mâle Accouche La Femelle," *Histoire de l'Académie Royale des Sciences* (1778): 13–19, https://www.biodiversitylibrary.org/item/89200#page/5/mode/1up; Schulte et al., "Developments in Amphibian."

over the next couple centuries: Schulte et al., "Developments in Amphibian."

Some frogs dug: Stefan K. Kaminsky, K. Eduard Linsenmair, and T. Ulmar Grafe, "Reproductive Timing, Nest Construction and Tadpole Guidance in the African Pig-Nosed Frog, *Hemisus Marmoratus*," *Journal of Herpetology* 33, no. 1 (March 1999): 119, https://doi.org/10.2307/1565550; Wells, *Ecology and Behavior*, 522, 530; Grafe, in discussion with the author, November 12, 2024.

Others whipped up foam: Fischer, "Form, Function, Foam"; Schulte, discussion.

Some frogs guarded eggs: Wells, *Ecology and Behavior*, 544–45; Ringler et al., "What Amphibians Can Teach"; Martha L. Crump, "Parental Care among the Amphibia," *Advances in the Study of Behavior*, vol. 25 (Elsevier, 1996), 109–44, https://doi.org/10.1016/S0065–3454(08)60331–9.

in one species in Chile: Oscar Goicoechea, Orlando Garrido, and Boris Jorquera, "Evidence for a Trophic Paternal-Larval Relationship in the Frog *Rhinoderma Darwinii*," *Journal of Herpetology* 20, no. 2 (June 1986): 168, https://doi.org/10.2307/1563941; Claudio Soto-Azat et al., "The Population Decline and Extinction of Darwin's Frogs," ed. Brian Gratwicke, *PLOS One* 8, no. 6 (June 12, 2013): e66957, https://doi.org/10.1371/journal.pone.0066957; "*Rhinoderma darwinii*," AmphibiaWeb, updated June 29, 2023, https://amphibiaweb.org/species/4322.

among frogs that looked after: Andrew I. Furness and Isabella Capellini, "The Evolution of Parental Care Diversity in Amphibians," *Nature Communications* 10, no. 1 (October 17, 2019): 4709, https://doi.org/10.1038/s41467-019-12608-5; Capellini, email message.

When biologists examined: This is based on a deeper dive into the supplementary data in Furness and Capellini, "Evolution of Parental Care"; Capellini, email message; Henshaw, email message to author, November 7, 2024.

in fish, males: Sigal Balshine, "Patterns of Parental Care in Vertebrates," in *The Evolution of Parental Care*, by Mathias Kölliker, eds. Nick J. Royle and Per T. Smiseth (Oxford University Press, 2012), 62–80, https://doi.org/10.1093/acprof:oso/9780199692576.003.0004.

that's what happens in mammals: Henshaw, discussion, February 1, 2024.

a science-fiction novel describing: Becky Chambers, *The Galaxy, and the Ground Within* (Harper Voyager, 2021), 208–11.

Borneo is home to: Tobias D. Jackson et al., "The Mechanical Stability of the World's Tallest Broadleaf Trees," *Biotropica* 53, no. 1 (January 2021): 110–20, https://doi.org/10.1111/btp.12850; Alexander Shenkin et al., "The World's Tallest Tropical Tree in Three Dimensions," *Frontiers in Forests and Global Change* 2 (June 18, 2019): 32, https://doi.org/10.3389/ffgc.2019.00032.

Many animals had evolved: Alfred Russel Wallace, *The Malay Archipelago* (London: Macmillan and Co., 1869), 59–61; John J. Socha, Tony O'Dempsey, and Michael LaBarbera, "A 3-D Kinematic Analysis of Gliding in a Flying Snake, *Chrysopelea Paradisi*," *Journal of Experimental Biology* 208, no. 10 (May 15, 2005): 1817–33, https://doi.org/10.1242/jeb.01579; John J. Socha, "Gliding Flight in *Chrysopelea*: Turning a Snake into a Wing," *Integrative and Comparative Biology* 51, no. 6 (December 2011): 969–82, https://doi.org/10.1093/icb/icr092; J. Maximilian Dehling, "How Lizards Fly: A Novel Type of Wing in Animals," *PLOS One* 12, no. 12 (December 13, 2017): e0189573, https://doi.org/10.1371/journal.pone.0189573; J. A. McGuire and R. Dudley, "The Biology of Gliding in Flying Lizards (Genus *Draco*) and Their Fossil and Extant Analogs," *Integrative and Comparative Biology* 51, no. 6 (December 1, 2011): 983–90, https://doi.org/10.1093/icb/icr090; Grafe, discussion, January 30, 2024; Chien Lee (biologist and wildlife photographer, Universiti Malaysia Sarawak), in discussion with the author, February 8, 2024.

Swooping, scurrying, and slithering: The descriptions in this paragraph were synthesized from Goyes Vallejos, in discussion with the author, November 20, 2018, February 6, 2024, March 8, 2022, November 14, 2023, and November 1, 2024; Grafe, discussion, January 30, 2024, and November 12, 2024; Hanyrol Ahmad Sah (herpetologist, Universiti Brunei Darussalam), in discussion with the author, April 18, 2024; Jennifer Sheridan (tropical conservation ecologist, Carnegie Museum of Natural History), in discussion with the author, January 30, 2024; "Rhinoceros Hornbill," Chien C. Lee Wildlife Photography, accessed March 24, 2025, https://photos.chienclee.com/image/I0000MIFAqU5PXNQ; "Horsfield's Tarsier," Chien C. Lee Wildlife Photography, accessed March 24, 2025, https://photos.chienclee.com/index/I0000QBcYiQxM8dg; "Leaf Rake-Like Antennae Beetle," Chien C. Lee Wildlife Photography, accessed March 24, 2025, https://photos.chienclee.com/index/I0000_YsLhPv2NMk; "Bornean Keeled Pit Viper," *Ecology Asia*, accessed March 24, 2025, https://www.ecologyasia.com/verts/snakes/bornean-keeled-pit-viper.htm; "Favolaschia Manipularis," Chien C. Lee Wildlife Photography, accessed March 24, 2025, https://photos.chienclee

.com/image/I0000H8hVouQWW20; Rhett Ayers Butler, "Glow-in-the-Dark Mushrooms in Borneo," *Mongabay*, November 11, 2012, https://news.mongabay.com/2012/11/glow-in-the-dark-mushrooms-in-borneo.

At night, wasps: Grafe, discussion, January 30, 2024; Xin Y. Er et al., "Venomous Stings and Bites in the Tropics (Malaysia): Review (Non-Snake Related)," *Open Access Library Journal* 8, no. 03 (2021): 1–24, https://doi.org/10.4236/oalib.1107230.

The biggest risk: Goyes Vallejos, in discussion with the author, March 8, 2022; Sheridan, discussion.

In most frog species: Wells, *Ecology and Behavior*, 278; Tumulty, discussion, October 30, 2024; Schulte, in discussion with the author, February 15, 2024, and November 6, 2024; Sarah L. Bush (behavioral ecologist, University of Missouri), in discussion with the author, February 9, 2024.

And sometimes, males grapple: Bush, discussion.

Early research on animal behavior: Henshaw, discussion, February 1, 2024; Hare, discussion.

Males compete: Robin M. Hare and Leigh W. Simmons, "Ecological Determinants of Sex Roles and Female Sexual Selection," in *Advances in the Study of Behavior*, 52 (2020): 1–28, https://doi.org/10.1016/bs.asb.2019.11.001; Charles Darwin, *The Descent of Man and Selection in Relation to Sex* (New York: D. Appleton and Company, 1871), 263–64, https://www.google.com/books/edition/The_Descent_of_man/ZvsHAAAAIAAJ?hl=en&gbpv=0; "Bird of Paradise," San Diego Zoo Wildlife Alliance, accessed March 24, 2025, https://animals.sandiegozoo.org/animals/bird-paradise-bird; Henshaw, discussion, November 7, 2024.

Scientists had observed the phenomenon: Fritzsche et al., "The 150th Anniversary."

First, consider a group: The information about seahorses and pipefishes was synthesized from Adam G. Jones and John C. Avise, "Male Pregnancy," *Current Biology* 13, no. 20 (2003): R791, https://doi.org/10.1016/j.cub.2003.09.045; Kimberly A. Paczolt and Adam G. Jones, "Post-Copulatory Sexual Selection and Sexual Conflict in the Evolution of Male Pregnancy," *Nature* 464, no. 7287 (March 2010): 401–4, https://doi.org/10.1038/nature08861; Fritzsche et al., "The 150th Anniversary"; Jones, discussion; Jones, email message to author, October 28, 2024; Hare, discussion.

Second, consider a species of: The information about bushcrickets was synthesized from Hare and Simmons, "Ecological Determinants of Sex Roles"; Hare, discussion; Hare, email message to author, October 30, 2024.

In both cases, the male: Fritzsche et al., "The 150th Anniversary"; Jones, discussion; Henshaw, discussion, February 1, 2024; Hare, discussion.

As a result, females: Fritzsche et al., "The 150th Anniversary"; Jones, discussion; Henshaw, discussion, February 1, 2024.

in one pipefish species: G. Rosenqvist and A. Berglund, "Sexual Signals and Mating Patterns in Syngnathidae," *Journal of Fish Biology* 78, no. 6 (June 2011): 1647–61, https://doi.org/10.1111/j.1095–8649.2011.02972.x; Anu Saikia, Jayanta Kumar Nath, and Dandadhar Sarma, "Observations on the Courtship Behaviour of Deocata Pipefish *Microphis Deocata* (Hamilton, 1822) (Actinopterygii: Syngnathiformes: Syngnathidae) in an Aquarium," *Journal of Threatened Taxa* 16, no. 1 (2024).

In certain dance flies: Rosalind L. Murray et al., "Sexual Selection on Multiple Female Ornaments in Dance Flies," *Proceedings of the Royal Society B: Biological Sciences* 285, no. 1887 (September 26, 2018): 20181525, https://doi.org/10.1098/rspb.2018.1525; David H. Funk and Douglas W. Tallamy, "Courtship Role Reversal and Deceptive Signals in the Long-Tailed Dance Fly, *Rhamphomyia Longicauda*," *Animal Behaviour* 59, no. 2 (February 2000): 411–21, https://doi.org/10.1006/anbe.1999.1310.

The mating roles seen: Fritzsche et al., "The 150th Anniversary."

the number of animals: Fritzsche et al., "The 150th Anniversary"; Hare and Simmons, "Ecological Determinants of Sex Roles"; Hare, discussion; Henshaw, discussion, November 7, 2024; "Summary Statistics," IUCN Red List, accessed March 24, 2025, https://www.iucnredlist.org/resources/summary-statistics#Summary%20Tables.

Scientists haven't found any frog: Jones, discussion; Summers, discussion; Bush, discussion.

in the green and black poison frog: Kyle Summers, "Sexual Conflict and Deception in Poison Frogs," *Current Zoology* 60, no. 1 (February 1, 2014): 37–42, https://doi.org/10.1093/czoolo/60.1.37; Tumulty, discussion, February 16, 2024; Summers, discussion.

biologists have started to question: Henshaw, discussion, February 1, 2024; Sara Lipshutz (behavioral ecologist, Duke University), in discussion with the author, March 7, 2024; Fischer, discussion.

Past research was heavily influenced: Malin Ah-King (evolutionary biologist and gender researcher, Örebro University), in discussion with the author, February 27, 2024; Malin Ah-King, "The History of Sexual Selection Research Provides Insights as to Why Females Are Still Understudied," *Nature Communications* 13, no. 1 (November 15, 2022): 6976, https://doi.org/10.1038/s41467-022-34770-z; Darwin, *Descent of Man*, 263–64.

While he did point out: Fritzsche et al., "The 150th Anniversary"; Hare, discussion; Ah-King, discussion.

Studies over the last few decades: Henshaw, discussion, February 1, 2024; Fischer, discussion.

Previous research on animal behavior: The information about bird mating was synthesized from Fischer, discussion; Lipshutz, discussion; Ah-King, discussion; Ah-King, "History of Sexual Selection Research"; Tim Birkhead, "Sperm Competition in Birds," *Trends in Ecology & Evolution* 2, no. 9 (September 1987): 268–72, https://doi.org/10.1016/0169–5347(87)90033–4; Susan M. Smith, "Extra-Pair Copulations in Black-Capped Chickadees: The Role of the Female," *Behaviour* 107, no. 1–2 (1988): 15–23, https://doi.org/10.1163/156853988X00160.

Bias can go: Ah-King, discussion.

some researchers think that scientists: Lipshutz, discussion; J. F. McLaughlin et al., "Multivariate Models of Animal Sex: Breaking Binaries Leads to a Better Understanding of Ecology and Evolution," *Integrative and Comparative Biology* 63, no. 4 (October 10, 2023): 891–906, https://doi.org/10.1093/icb/icad027.

Among humans, the idea: Sherry Jueyu Wu and Xiqian Cai, "Adding Up Peer Beliefs: Experimental and Field Evidence on the Effect of Peer Influence on Math Performance," *Psychological Science* 34, no. 8 (August 2023): 851–62, https://doi.org/10.1177/09567976231180881.

On her third expedition: The results of this study are described in Goyes Vallejos, Grafe, and Wells, "Prolonged Parental Behaviour."

male coquí frogs sat: Daniel S. Townsend, Margaret M. Stewart, and F. Harvey Pough, "Male Parental Care and Its Adaptive Significance in a Neotropical Frog," *Animal Behaviour* 32, no. 2 (May 1984): 421–31, https://doi.org/10.1016/S0003–3472(84)80278-X; Daniel S. Townsend, "The Costs of Male Parental Care and Its Evolution in a Neotropical Frog," *Behavioral Ecology and Sociobiology* 19, no. 3 (August 1986): 187–95, https://doi.org/10.1007/BF00300859; Margaret M. Stewart and A. Stanley Rand, "Vocalizations and the Defense of Retreat Sites by Male and Female Frogs, *Eleutherodactylus Coqui*," *Copeia* 1991, no. 4 (December 13, 1991): 1013, https://doi.org/10.2307/1446096; Sarah E. Westrick, Mara Laslo, and Eva K. Fischer, "The Big Potential of the Small Frog *Eleutherodactylus Coqui*," *eLife* 11 (January 14, 2022): e73401, https://doi.org/10.7554/eLife.73401.

In Majorcan midwife toads: Sarah L. Bush and Diana J. Bell, "Courtship and Female Competition in the Majorcan Midwife Toad, *Alytes Muletensis*," *Ethology* 103, no. 4 (April 1997): 292–303, https://doi.org/10.1111/j.1439–0310.1997.tb00019.x; Bush, discussion.

But Johana would need: Jones, discussion; Hare, discussion.

a viral disease: Olivia Z. Daniel et al., "Rapid Spread of African Swine Fever Across Borneo," bioRxiv (2024), https://doi.org/10.1101/2024.06.20.597708;

Kymberley Chu, SL Wong, and Alven Chang, "Where Are All the Sabah Pigs?," *Macaranga*, February 16, 2022, https://www.macaranga.org/where-are-all-the-sabah-pigs; Lee, discussion; Joremy anak Tony (research assistant, Universiti Brunei Darussalam), in discussion with the author, April 29, 2024.

In the past, the pigs: Goyes Vallejos, discussion, February 26, 2024; Goyes Vallejos, Grafe, and Wells, "Factors Influencing Tadpole Deposition."

Bornean horned frogs honked: Ulmar Grafe, "*Megophrys Nasuta* (Bornean Horned Frog)," *Frog Voices of Borneo*, accessed March 24, 2025, https://soundcloud.com/frogvoicesofborneo/megophrys-nasuta-bornean-horned-frog.

Astronomy: Moon on Fire

Details about the WASP-49 b research were drawn from interviews with team members Apurva Oza, Julia Seidel, Jens Hoeijmakers, Athira Unni, Andrea Gebek, Bob Johnson, Aurora Kesseli, Carl Schmidt, Rosaly Lopes, or Ashley Baker; from team Zoom meetings that I attended; or from the scientific papers Apurva V. Oza et al., "Sodium and Potassium Signatures of Volcanic Satellites Orbiting Close-in Gas Giant Exoplanets," *The Astrophysical Journal* 885, no. 2 (November 12, 2019): 168, https://doi.org/10.3847/1538–4357/ab40cc; Andrea Gebek and Apurva V. Oza, "Alkaline Exospheres of Exoplanet Systems: Evaporative Transmission Spectra," *Monthly Notices of the Royal Astronomical Society* 497, no. 4 (October 1, 2020): 5271–91, https://doi.org/10.1093/mnras/staa2193; and Apurva V. Oza et al., "Redshifted Sodium Transient Near Exoplanet Transit," *The Astrophysical Journal Letters* 973, no. 2 (October 1, 2024): L53, https://doi.org/10.3847/2041–8213/ad6b29. Details about exomoon research by David Kipping and his colleagues were drawn from interviews with Kipping and Alex Teachey, as well as the team's numerous papers and videos on the team's *Cool Worlds* YouTube channel (specific citations are listed below). Biographical details about Oza were drawn from interviews with him. General background about exomoon research was drawn from interviews with team members and other experts such as Sébastien Charnoz, Michelle Hill, Cecilia Lazzoni, Matthew Penny, Andrew Vanderburg, Mary Anne Limbach, and Melinda Soares-Furtado. Details about Jupiter and Io were drawn from interviews with Johnson, Oza, and Katherine de Kleer.

This chapter was meant to capture a particular period in the middle of the team's research, from 2021 to early 2023, but Oza and his colleagues continued to gather more evidence afterward. For details about what they found in later analyses, see Oza et al., "Redshifted Sodium Transient"; "Does Distant Planet Host Volcanic Moon Like Jupiter's Io?," NASA, October 10, 2024, https://www.nasa.gov/universe/exoplanets/does-distant-planet-host-volcanic-moon-like-jupiters-io; Athira Unni et al., "Doppler Shifted Transient Sodium Detection by KECK/HIRES," *Monthly*

Notices of the Royal Astronomical Society: Letters 540, no. 1 (2025): L48–53, https://doi.org/10.1093/mnrasl/slaf031.

One moon, Io: "Moons of Jupiter," NASA, accessed March 25, 2025, https://science.nasa.gov/jupiter/jupiter-moons.

When Galileo had first seen: Galileo Galilei, *Sidereus Nuncius or The Sidereal Messenger*, trans. Albert Van Helden, 2nd ed. (1610; repr., University of Chicago Press, 2015), 66.

Scientists had started spitballing: P. Sartoretti and J. Schneider, "On the Detection of Satellites of Extrasolar Planets with the Method of Transits," *Astronomy and Astrophysics Supplement Series* 134, no. 3 (February 1999): 553–60, https://doi.org/10.1051/aas:1999148.

David Kipping, an astronomer: Alex Teachey and David M. Kipping, "Evidence for a Large Exomoon Orbiting Kepler-1625b," *Science Advances* 4, no. 10 (October 2018): eaav1784, https://doi.org/10.1126/sciadv.aav1784; David Kipping et al., "An Exomoon Survey of 70 Cool Giant Exoplanets and the New Candidate Kepler-1708 b-i," *Nature Astronomy* 6, no. 3 (January 13, 2022): 367–80, https://doi.org/10.1038/s41550-021-01539-1.

Other researchers had turned up: C. Lazzoni et al., "The Search for Disks or Planetary Objects around Directly Imaged Companions: A Candidate around DH Tauri B," *Astronomy & Astrophysics* 641 (September 2020): A131, https://doi.org/10.1051/0004–6361/201937290; D. P. Bennett et al., "MOA-2011-BLG-262Lb: A Sub-Earth-Mass Moon Orbiting a Gas Giant Primary or a High Velocity Planetary System in the Galactic Bulge," *The Astrophysical Journal* 785, no. 2 (April 7, 2014): 155, https://doi.org/10.1088/0004–637X/785/2/155.

But the scientific community: Alex Teachey, "Detecting and Characterizing Exomoons and Exorings (*Handbook of Exoplanets*, 2nd Edition)," arXiv:2401.13293 (January 24, 2024), https://doi.org/10.48550/arXiv.2401.13293; René Heller and Michael Hippke, "Large Exomoons Unlikely Around Kepler-1625 b and Kepler-1708 b," *Nature Astronomy* 8, no. 2 (December 7, 2023): 193–206, https://doi.org/10.1038/s41550-023-02148-w; Sébastien Charnoz (astrophysicist, Université Paris Cité), in discussion with the author, June 9, 2022.

And numerous other cases: David Kipping (astronomer, Columbia University), in discussion with the author, September 16, 2022; Teachey, "Detecting and Characterizing Exomoons"; David Kipping, "An Independent Analysis of the Six Recently Claimed Exomoon Candidates," *The Astrophysical Journal* 900, no. 2 (September 15, 2020): L44, https://doi.org/10.3847/2041–8213/abafa9.

Typical news headlines: Govert Shilling, "Have Astronomers Detected Exomoons at Last?," *Sky & Telescope*, August 29, 2019, https://skyandtelescope.org/astronomy-news/have-astronomers-detected-exomoons; Sabina Sagynbayeva, "Astronomers Have Spotted 6 New Moons in Other Planetary Systems! Or Have They?," *Astrobites*, January 25, 2021, https://astrobites.org/2021/01/25/6-new-moons.

Exomoons hold fascination: Teachey, "Detecting and Characterizing Exomoons."

One of NASA's goals: "Kepler / K2," NASA, accessed March 25, 2025, https://science.nasa.gov/mission/kepler; "Transiting Exoplanet Survey Satellite (TESS)," NASA, accessed March 25, 2025, https://exoplanets.nasa.gov/tess.

A moon has extra sources: Michelle Hill (exoplanet astronomer, Stanford University), in discussion with the author, April 13, 2021.

romantic sci-fi visions: "Ewok," *Star Wars*, accessed March 25, 2025, https://www.starwars.com/databank/ewok; Carol Kaesuk Yoon, "Luminous 3-D Jungle Is a Biologist's Dream," *New York Times*, January 18, 2010, https://www.nytimes.com/2010/01/19/science/19essay.html.

The moons he sought: "Hints of a Volcanically Active Exomoon," University of Bern, August 29, 2019, https://mediarelations.unibe.ch/media_releases/2019/medienmitteilungen_2019/hints_of_a_volcanically_active_exomoon/index_eng.html.

He still leaned: "Hints of a Volcanically."

finding the first exomoon: Alex Teachey (astronomer), in discussion with the author, April 17, 2023; Hill, discussion, April 18, 2023; "In Depth: Exoplanets," NASA, accessed March 25, 2025, https://science.nasa.gov/exoplanets/facts.

The hunt for exoplanets: Joshua Winn, "Who Really Discovered the First Exoplanet?," *Scientific American* (blog), November 12, 2019, https://www.scientificamerican.com/blog/observations/who-really-discovered-the-first-exoplanet.

A pair of researchers: "Nobel Prize in Physics 2019," The Nobel Prize, accessed March 25, 2025, https://www.nobelprize.org/prizes/physics/2019/summary. Whether they deserve this honor over other exoplanet scientists is debatable. In Winn, "Who Really Discovered," astronomer Joshua Winn argues that several other teams were also "credible contenders" for the title.

Thousands of exoplanets: "How Many Exoplanets Are There?," NASA, accessed March 25, 2025, https://science.nasa.gov/exoplanets/how-many-exoplanets-are-there; "Exoplanets," NASA, accessed March 25, 2025, https://science.nasa.gov/exoplanets.

Teachey once described: This quote is from an early unpublished draft of the paper Teachey, "Detecting and Characterizing Exomoons."

a type of exoplanet: Jonathan J. Fortney, Rebekah I. Dawson, and Thaddeus D. Komacek, "Hot Jupiters: Origins, Structure, Atmospheres," *Journal of Geophysical Research: Planets* 126, no. 3 (March 2021): e2020JE006629, https://doi.org/10.1029/2020JE006629; Hill, in discussion with the fact-checker, October 16, 2024.

I'd read an account: Linda A. Morabito, "Discovery of Volcanic Activity on Io: A Historical Review," arXiv:1211.2554 (November 12, 2012), https://doi.org/10.48550/arXiv.1211.2554.

But "the more you work": Morabito, "Discovery of Volcanic Activity," 25.

coming across a paper: Oza et al., "Sodium and Potassium Signatures."

About 3,700 trillion miles: "WASP-49 b," NASA, last modified October 24, 2024, https://science.nasa.gov/exoplanet-catalog/wasp-49-b; Apurva Oza (planetary astrophysicist, Jet Propulsion Laboratory), in discussion with the author, December 15, 2022; "Luminous in Lepus," European Space Agency, accessed March 25, 2025, https://www.esa.int/ESA_Multimedia/Images/2024/01/Luminous_in_Lepus.

he described an image: The deep azure in that image is not Neptune's true color. A 2024 study suggested that the planet is actually a much paler bluish green: Patrick G. J. Irwin et al., "Modelling the Seasonal Cycle of Uranus's Colour and Magnitude, and Comparison with Neptune," *Monthly Notices of the Royal Astronomical Society* 527, no. 4 (2024): 11521–38, https://doi.org/10.1093/mnras/stad3761.

seeing that same shade: An example of Klein's blue paintings can be seen at "Yves Klein," Tate Gallery, accessed March 25, 2025, https://www.tate.org.uk/art/artists/yves-klein-1418.

picture taken by the Hubble: "Hubble's Deep Fields," NASA, accessed March 25, 2025, https://science.nasa.gov/mission/hubble/science/universe-uncovered/hubble-deep-fields.

posted a 1969 article: "Deux Hommes Ont Foulé le Sol de la Lune Devant des Centaines de Millions de Téléspectateurs," *Le Monde*, July 22, 1969, https://www.lemonde.fr/archives/article/1969/07/22/deux-hommes-ont-foule-le-sol-de-la-lune-devant-des-centaines-de-millions-de-telespectateurs-oui-mais-pourquoi_3058704_1819218.html.

he'd created a Dance: "Dance Your PhD 2017: O2 'Panga' on Europa," Oza ***, September 29, 2017, YouTube video, 8:35, https://www.youtube.com/watch?v=MVMgEvBf_sw.

Robert and a colleague: R. E. Johnson and P. J. Huggins, "Toroidal Atmospheres Around Extrasolar Planets," *Publications of the Astronomical Society of the Pacific* 118, no. 846 (August 2006): 1136–43, https://doi.org/10

.1086/506183; Robert Johnson (molecular physicist, University of Virginia), in discussion with the author, February 23, 2022.

On March 2, 1979: The information about the Io prediction and gravitational forces was synthesized from S. J. Peale, P. Cassen, and R. T. Reynolds, "Melting of Io by Tidal Dissipation," *Science* 203, no. 4383 (March 2, 1979): 892–94, https://doi.org/10.1126/science.203.4383.892; Dale P. Cruikshank and Robert M. Nelson, "A History of the Exploration of Io," in *Io After Galileo*, eds. Rosaly M. C. Lopes and John R. Spencer (Springer Berlin Heidelberg, 2007), 5–33; Katherine de Kleer (planetary astronomer, California Institute of Technology), in discussion with the author, March 2, 2022; Oza, discussion, April 21, 2021; Hill, discussion, April 13, 2021; Johnson, in discussion with the author, May 6, 2025.

"one might speculate that widespread": Peale, Cassen, and Reynolds, "Melting of Io by Tidal Dissipation," 894.

the spacecraft *Voyager 1*: Bradford A. Smith et al., "The Jupiter System Through the Eyes of Voyager 1," *Science* 204, no. 4396 (June 1979): 951–72, https://doi.org/10.1126/science.204.4396.951; Morabito, "Discovery of Volcanic Activity."

She worked in an office: Morabito, "Discovery of Volcanic Activity," 7.

"Jesus, what's that?": Morabito, "Discovery of Volcanic Activity," 14.

The colleague had heard: The information in the first part of this paragraph was synthesized from Morabito, "Discovery of Volcanic Activity"; L. A. Morabito et al., "Discovery of Currently Active Extraterrestrial Volcanism," *Science* 204, no. 4396 (1979): 972; Robert G. Strom et al., "Volcanic Eruption Plumes on Io," *Nature* 280, no. 5725 (August 1979): 733–36, https://doi.org/10.1038/280733a0; Iain Todd, "Interview with Voyager Scientist Linda Morabito," *BBC Sky at Night Magazine*, July 6, 2020, https://www.skyatnightmagazine.com/space-missions/volcano-jupiter-moon-io-interview-voyager-linda-morabito; A. J. S. Rayl, "The Stories Behind the Voyager Mission: Linda Morabito Kelly," *The Planetary Society*, September 5, 2002, https://www.planetary.org/articles/stories_kelly; John Noble Wilford, "Voyager Finds Io Alive with Volcanoes," *New York Times*, March 13, 1979, https://www.nytimes.com/1979/03/13/archives/voyager-finds-io-alive-with-volcanoes-active-volcanoes-found-on-io.html.

On March 12: Morabito, "Discovery of Volcanic Activity," 28.

They had found the first: Morabito et al., "Discovery of Currently Active."

The semicircle was: Smith et al., "The Jupiter System," 961.

one element released: Johnson and Huggins, "Toroidal Atmospheres"; Bruce Fegley and Mikhail Zolotov, "Chemistry of Sodium, Potassium, and Chlorine in Volcanic Gases on Io," *Icarus* 148, no. 1 (November 2000): 193–210, https://doi.org/10.1006/icar.2000.6490.

In 1972, an astronomer: Robert A. Brown, "Optical Line Emission from Io," in *Exploration of the Planetary System*, ed. A. Woszczyk and C. Iwaniszewska (Springer, 1974), 527.

scientists cataloged even more: Michael Mendillo et al., "The Extended Sodium Nebula of Jupiter," *Nature* 348, no. 6299 (November 22, 1990): 312–14, https://doi.org/10.1038/348312a0; Nicholas M. Schneider et al., "Molecular Origin of Io's Fast Sodium," *Science* 253, no. 5026 (1991): 1394–97; E. Lellouch et al., "Volcanically Emitted Sodium Chloride as a Source for Io's Neutral Clouds and Plasma Torus," *Nature* 421, no. 6918 (January 2003): 45–47, https://doi.org/10.1038/nature01292. For a review, see Nicholas M. Schneider, "Io's Escaping Atmosphere: Continuing the Legacy of Surprise," *Proceedings of the International Astronomical Union* 6, no. S269 (January 2010): 80–86, https://doi.org/10.1017/S1743921310007295.

Then, in 2002, astronomers reported: David Charbonneau et al., "Detection of an Extrasolar Planet Atmosphere," *The Astrophysical Journal* 568, no. 1 (March 20, 2002): 377–84, https://doi.org/10.1086/338770; "HD 209458b," NASA, accessed March 26, 2025, https://science.nasa.gov/resource/hd-209458b.

Those researchers assumed: Charbonneau et al., "Detection of an Extrasolar."

In a 2006 paper, Robert: Johnson and Huggins, "Toroidal Atmospheres"; Johnson, discussion.

Sodium happens to be easy: Carl Schmidt (planetary scientist, Boston University), in discussion with the author, March 7, 2022; Johnson, discussion, May 6, 2025.

eloquently described in Morabito's account: Morabito, "Discovery of Volcanic Activity," 16.

Around the same time: Teachey, "Detecting and Characterizing Exomoons."

He'd done some theoretical work: David M. Kipping, "Transit Timing Effects Due to an Exomoon," *Monthly Notices of the Royal Astronomical Society* 392, no. 1 (January 1, 2009): 181–89, https://doi.org/10.1111/j.1365-2966.2008.13999.x; David M. Kipping, "Transit Timing Effects Due to an Exomoon—II," *Monthly Notices of the Royal Astronomical Society* 396, no. 3 (July 1, 2009): 1797–1804, https://doi.org/10.1111/j.1365-2966.2009.14869.x; David M. Kipping, Stephen J. Fossey, and Giammarco Campanella, "On the Detectability of Habitable Exomoons with Kepler-Class Photometry," *Monthly Notices of the Royal Astronomical Society* 400, no. 1 (November 21, 2009): 398–405, https://doi.org/10.1111/j.1365-2966.2009.15472.x.

***Popular Science* picked him:** Veronique Greenwood and Cassandra

Willyard, "Brilliant 10: David Kipping Hunts for Moons Around Exoplanets," *Popular Science*, archived August 3, 2024, https://web.archive.org/web/20240803155127/https://www.popsci.com/brilliant-10-david-kipping-hunts-for-moons-around-exoplanets.

On his lab's YouTube channel: "We Discovered a New Exomoon Candidate!!," *Cool Worlds*, January 13, 2022, YouTube video, 29:16, https://www.youtube.com/watch?v=Blej3YvveCI.

Other astronomers had speculated: The details about the transit method in these two paragraphs were synthesized from Sartoretti and Schneider, "On the Detection of Satellites"; Teachey, "Detecting and Characterizing Exomoons"; Kipping, in discussion with the author, 2013; Teachey, in discussion with the author, March 15, 2022; Vanderburg, discussion; "What's a Transit?," NASA, accessed March 26, 2025, https://science.nasa.gov/exoplanets/whats-a-transit.

he ticked off a long list: The explanation of transit method challenges is primarily from Charnoz, discussion. To explain these concepts, I also drew on Teachey, discussion, March 15, 2022; Vanderburg, discussion; Mary Anne Limbach (astrophysicist, University of Michigan), in discussion with the author, June 8, 2022; Julia Seidel (exoplanet astronomer, European Southern Observatory), in discussion with the author, March 25, 2022.

The seeming impossibility: Depending on which source you consult, his full name was Guillaume-Joseph-Hyacinthe-Jean-Baptiste Le Gentil, Guillaume-Hyacinthe-Joseph-Jean-Baptiste Le Gentil de La Galaisière, or some other variation on that theme. The information about Le Gentil was synthesized from Helen Sawyer Hogg, "Out of Old Books: Le Gentil and the Transits of Venus, 1761 and 1769," *Journal of the Royal Astronomical Society of Canada*, 45 (February 1951): 37–46, https://ui.adsabs.harvard.edu/abs/1951JRASC..45...37S/abstract; Hogg, "Out of Old Books," 45 (April 1951): 89–92, https://ui.adsabs.harvard.edu/abs/1951JRASC..45...89S/abstract; Hogg, "Out of Old Books," 45 (June 1951): 127–34, https://ui.adsabs.harvard.edu/abs/1951JRASC..45..127S/abstract; Hogg, "Out of Old Books," 45 (August 1951): 173–78, https://ui.adsabs.harvard.edu/abs/1951JRASC..45..173S/abstract; Francine Jackson, "Le Gentil de la Galaisière, Guillaume-Joseph-Hyacinthe Jean-Baptiste," in *Biographical Encyclopedia of Astronomers*, eds. Thomas Hockey et al. (New York: Springer, 2014), 1296–97; Jean-Dominique Cassini, *Mémoires Pour Servir A L'Histoire des Science et a Celle de L'Observatoire Royal De Paris* (Paris: Bleuet, 1810), 358; "Voyage Dans les Mers de l'Inde," Bibliothèque Nationale De France, accessed March 26, 2025, https://gallica.bnf.fr/ark:/12148/bpt6k6244383b; Eli Maor, *Venus in Transit* (Princeton, New Jersey: Princeton University Press, 2004), 85. Additional details about the transits of Venus were drawn from Fred Espenak, "2012

Transit of Venus," in *RASC Observer's Handbook* (Royal Astronomical Society of Canada, 2012), 144–52, and "Venus Transit Across the Sun (2014)," NASA, accessed March 26, 2025, https://science.nasa.gov/resource/venus-transit-across-the-sun-2014. Le Gentil's quotes are from Hogg, "Out of Old Books" (June 1951): 132.

In 2009, NASA's Kepler: "Kepler / K2"; Natalie M. Batalha et al., "Planetary Candidates Observed by Kepler. III. Analysis of the First 16 Months of Data," *The Astrophysical Journal Supplement Series* 204, no. 24 (2013): 1–21.

Kipping, who had moved: David Nesvorný et al., "The Detection and Characterization of a Nontransiting Planet by Transit Timing Variations," *Science* 336, no. 6085 (June 2012): 1133–36, https://doi.org/10.1126/science.1221141; Kipping, discussion, September 16, 2022.

Kipping kept looking: David M. Kipping et al., "The Hunt for Exomoons with Kepler (HEK). V. A Survey of 41 Planetary Candidates for Exomoons," *The Astrophysical Journal* 813, no. 1 (October 22, 2015): 14, https://doi.org/10.1088/0004–637X/813/1/14 (see discussion of PH-2b on 7, 10); Kipping, discussion, September 16, 2022.

His hopes were dashed again: J. Cabrera et al., "The Planetary System to KIC 11442793: A Compact Analogue to the Solar System," *The Astrophysical Journal* 781, no. 1 (2014): 18, https://doi.org/10.1088/0004–637X/781/1/18; "Discovery of Eight Planets Makes Alien System the First to Tie with Our Solar System," NASA, December 14, 2017, https://science.nasa.gov/universe/exoplanets/discovery-of-eight-planets-makes-alien-system-the-first-to-tie-with-our-solar-system.

The signal turned out to be: David M. Kipping et al., "The Possible Moon of Kepler-90g Is a False Positive," *The Astrophysical Journal* 799, no. 1 (January 20, 2015): L14, https://doi.org/10.1088/2041–8205/799/1/L14; Kipping, discussion, September 16, 2022; Jessie L. Christiansen et al., "Measuring Transit Signal Recovery in the Kepler Pipeline. I. Individual Events," *The Astrophysical Journal Supplement Series* 207 (2013): 35, https://doi.org/10.1088/0067–0049/207/2/35.

Like other exoplanet astronomers: The information about transit spectroscopy was synthesized from Ashley Baker (astronomer, California Institute of Technology), in discussion with the author, February 7, 2022; "Spectroscopy Infographic," NASA, accessed March 27, 2025, https://science.nasa.gov/resource/spectroscopy-infographic; "Spectroscopy—Detection of Biosignatures," NASA, accessed March 27, 2025, https://science.nasa.gov/resource/spectroscopy-detection-of-biosignatures; Oza et al., "Sodium and Potassium Signatures"; Francesco Pepe et al., "ESPRESSO—An Echelle SPectrograph for Rocky Exoplanets Search and Stable Spectroscopic

Observations," *The Messenger* 153 (2013): 6–16; Seidel, in discussion with the author, February 17, 2022.

In 2017, a new report: Information about Wyttenbach's study, WASP-49 b, and sodium absorption was synthesized from Aurélien Wyttenbach et al., "Hot Exoplanet Atmospheres Resolved with Transit Spectroscopy (HEARTS): I. Detection of Hot Neutral Sodium at High Altitudes on WASP-49b," *Astronomy & Astrophysics* 602 (June 2017): A36, https://doi.org/10.1051/0004–6361/201630063; Aurélien Wyttenbach et al., "Spectrally Resolved Detection of Sodium in the Atmosphere of HD 189733b with the HARPS Spectrograph," *Astronomy & Astrophysics* 577 (May 2015): A62, https://doi.org/10.1051/0004–6361/201525729; Oza, in discussion with the author, June 8, 2021, and February 3, 2022; Seidel, discussion, February 17, 2022; Jens Hoeijmakers (exoplanet astronomer, Lund University), in discussion with the author, February 16, 2022; Baker, discussion; "WASP-49 b," NASA; Oza et al., "Redshifted Sodium Transient."

Wyttenbach's team had concluded: Details about differing interpretations of the sodium signal were synthesized from Wyttenbach et al., "Hot Exoplanet Atmospheres"; Oza et al., "Sodium and Potassium Signatures"; Oza, discussion, April 21, 2021, and February 3, 2022; Hoeijmakers, discussion, February 16, 2022; Andrea Gebek (PhD student in astrophysics, Ghent University), in discussion with the author, February 17, 2022.

On a December night: Details about the Very Large Telescope were synthesized from Jonathan Smoker (astronomer, Very Large Telescope), in discussion with the author, March 14, 2022; Oza, in discussion with the author, February 28, 2022; "Paranal Observatory," European Southern Observatory, accessed March 27, 2025, https://www.eso.org/public/teles-instr/paranal-observatory; "La Silla Paranal Observatory: The Paranal Facilities," European Southern Observatory, accessed March 27, 2025, https://www.eso.org/sci/facilities/paranal.html; "A Giant of Astronomy and a Quantum of Solace," European Southern Observatory, March 25, 2008, https://www.eso.org/public/news/eso0807; "Bond@Paranal," European Southern Observatory, accessed March 27, 2025, https://www.eso.org/public/archives/static/events/special-evt/bond/BondatParanal.html; "Very Large Telescope," European Southern Observatory, accessed March 27, 2025, https://www.eso.org/public/teles-instr/paranal-observatory/vlt; Oza et al., "Redshifted Sodium Transient."

In the modern era: Smoker, discussion; Oza, in discussion with the author, May 26, 2021.

ESPRESSO was the name: Pepe et al., "ESPRESSO."

another telescope called UT3: UT3 is the technical designation. In 1999, the telescope was named Melipal, meaning "The Southern Cross": "Names of VLT

Unit Telescopes," European Southern Observatory, accessed March 27, 2025, https://www.eso.org/public/teles-instr/paranal-observatory/vlt/vlt-names; "Melipal," European Southern Observatory, accessed March 27, 2025, https://www.eso.org/public/teles-instr/paranal-observatory/vlt/vlt-names/melipal.

he and Teachey had tried: The description of the Kepler-1625b analysis was synthesized from Teachey, discussion, March 15, 2022; Kipping, discussion, September 16, 2022; Teachey and Kipping, "Evidence for a Large Exomoon"; A. Teachey, D. M. Kipping, and A. R. Schmitt, "HEK. VI. On the Dearth of Galilean Analogs in Kepler, and the Exomoon Candidate Kepler-1625b I," *The Astronomical Journal* 155, no. 1 (2018): 36, https://doi.org/10.3847/1538–3881/aa93f2; Teachey, "Detecting and Characterizing Exomoons."

The moon, if it existed: "Did We Find an Exomoon?," *Cool Worlds*, July 27, 2017, YouTube video, 13:27, https://www.youtube.com/watch?v=8V9QHn5oHMs; "We Discovered A New Exomoon Candidate!!"

Based on the size: Teachey and Kipping, "Evidence for a Large Exomoon"; "We Discovered a New Exomoon Candidate!!"

Kipping and Teachey presented: Teachey and Kipping, "Evidence for a Large Exomoon," 7.

Two teams reanalyzed: Laura Kreidberg, Rodrigo Luger, and Megan Bedell, "No Evidence for Lunar Transit in New Analysis of *Hubble Space Telescope* Observations of the Kepler-1625 System," *The Astrophysical Journal* 877, no. 2 (May 24, 2019): L15, https://doi.org/10.3847/2041–8213/ab20c8; René Heller, Kai Rodenbeck, and Giovanni Bruno, "An Alternative Interpretation of the Exomoon Candidate Signal in the Combined *Kepler* and *Hubble* Data of Kepler-1625," *Astronomy & Astrophysics* 624 (April 2019): A95, https://doi.org/10.1051/0004–6361/201834913.

One team thought: Heller, Rodenbeck, and Bruno, "An Alternative Interpretation," 8.

The second team: Kreidberg, Luger, and Bedell, "No Evidence for Lunar Transit," 1.

By this time, other teams: Teachey, "Detecting and Characterizing Exomoons."

There were gaps: M. A. Kenworthy and E. E. Mamajek, "Modeling Giant Extrasolar Ring Systems in Eclipse and the Case of J1407b: Sculpting by Exomoons?," *The Astrophysical Journal* 800, no. 2 (February 18, 2015): 126, https://doi.org/10.1088/0004–637X/800/2/126.

There was a candidate: Bennett et al., "MOA-2011-BLG-262Lb." The explanation of microlensing was synthesized from Matthew Penny (astronomer, Louisiana State University), in discussion with the author, July 8, 2022; Teachey, discussion, March 15, 2022; Hill, discussion, March 15, 2022.

There was a blob: Lazzoni et al., "The Search for Disks"; Cecilia Lazzoni (exoplanet scientist, Istituto Nazionale di Astrofisica), in discussion with the author, March 15, 2022.

But all of these candidates: Teachey, "Detecting and Characterizing Exomoons"; Teachey, discussion, March 15, 2022; Penny, discussion; Lazzoni, discussion.

The upshot, as Kipping: "We Discovered a New Exomoon Candidate!!"

Like Kipping, the team faced: Seidel, discussion, March 25, 2022.

Finally, they needed to address: N. Casasayas-Barris et al., "Is There Na I in the Atmosphere of HD 209458b?: Effect of the Centre-to-Limb Variation and Rossiter-McLaughlin Effect in Transmission Spectroscopy Studies," *Astronomy & Astrophysics* 635 (March 2020): A206, https://doi.org/10.1051/0004-6361/201937221; Seidel, discussion, February 17, 2022, and March 25, 2022.

Wyttenbach's team had concluded: Wyttenbach et al., "Hot Exoplanet Atmospheres."

his team was examining "cool giants": The description of the Kepler 1708 b analysis was synthesized from Kipping et al., "An Exomoon Survey"; Kipping, discussion, September 16, 2022; Teachey, discussion, March 15, 2022; "We Discovered a New Exomoon Candidate!!"

He told me about an astronomer: Steven S. Vogt et al., "The Lick-Carnegie Exoplanet Survey: A 3.1 M⊕ Planet in the Habitable Zone of the Nearby M3V Star Gliese 581," *The Astrophysical Journal* 723, no. 1 (November 1, 2010): 954–65, https://doi.org/10.1088/0004-637X/723/1/954; Yudhijit Bhattacharjee, "Data Dispute Revives Exoplanet Claim," *Science* 337 (2012): 398.

Later research suggested the planet: Rene Andrae, Tim Schulze-Hartung, and Peter Melchior, "Dos and Don'ts of Reduced Chi-Squared," arXiv (December 16, 2010), http://arxiv.org/abs/1012.3754; T. Forveille et al., "The HARPS Search for Southern Extra-Solar Planets XXXII. Only 4 Planets in the Gl 581 System," arXiv (September 12, 2011), http://arxiv.org/abs/1109.2505; Philip C. Gregory, "Bayesian Re-Analysis of the Gliese 581 Exoplanet System: Bayesian Re-Analysis of Gliese 581," *Monthly Notices of the Royal Astronomical Society* 415, no. 3 (August 11, 2011): 2523–45, https://doi.org/10.1111/j.1365-2966.2011.18877.x; R. V. Baluev, "The Impact of Red Noise in Radial Velocity Planet Searches: Only Three Planets Orbiting GJ 581?," *Monthly Notices of the Royal Astronomical Society* 429, no. 3 (March 1, 2013): 2052–68, https://doi.org/10.1093/mnras/sts476; A. P. Hatzes, "An Investigation into the Radial Velocity Variability of GJ 581: On the Significance of GJ 581g," *Astronomische Nachrichten* 334, no. 7 (2013): 616–24; Paul Robertson et al., "Stellar Activity Masquerading as Planets in the Habitable Zone of the M Dwarf Gliese 581," *Science* 345, no. 6195 (July 25, 2014): 440–44, https://

doi.org/10.1126/science.1253253; Abel Méndez, "Habitable Exoplanets," in *New Frontiers in Astrobiology* (Elsevier, 2022), 179–92, https://doi.org/10.1016/B978-0-12-824162-2.00011-7. Vogt's responses can be found in S. S. Vogt, R. P. Butler, and N. Haghighipour, "GJ 581 Update: Additional Evidence for a Super-Earth in the Habitable Zone," *Astronomische Nachrichten* 333, no. 7 (August 2012): 561–75, https://doi.org/10.1002/asna.201211707; Mike Wall, "R.I.P. Possibly Habitable Planet Gliese 581g? Not So Fast, Co-Discoverer Says," *Space.com*, February 19, 2011, https://www.space.com/10897-alien-planet-gliese-581g-great-debate.html; Bhattacharjee, "Data Dispute"; Dennis Overbye, "A Planet 'Just Right' for Life? Perhaps, if It Exists," *New York Times*, August 20, 2012, https://www.nytimes.com/2012/08/21/science/space/just-right-or-nonexistent-dispute-over-goldilocks-planet-gliese-581g.html.

a team in Germany: René Heller and Michael Hippke, "Large Exomoons Unlikely Around Kepler-1625 b and Kepler-1708 b," *Nature Astronomy* 8, no. 2 (December 7, 2023): 193–206, https://doi.org/10.1038/s41550-023-02148-w; "Giant Doubts About Giant Exomoons," Max-Planck-Gesellschaft, December 8, 2023, https://www.mpg.de/21217437/1205-aero-giant-doubts-about-giant-exomoons-151060-x.

Kipping's team then wrote: David Kipping et al., "A Reply to: Large Exomoons Unlikely Around Kepler-1625 b and Kepler-1708 b," arXiv (January 18, 2024), 7, http://arxiv.org/abs/2401.10333.

One news story ran: The story's dramatic graphic doesn't appear to be online anymore or available via the *Wayback Machine*, but the text of the story can be found at Evan Gough, "Exomoons Defy Discovery," *Universe Today*, December 11, 2023, https://www.universetoday.com/articles/exomoons-defy-discovery.

Wyttenbach's team had averaged: The description of the team's reanalysis was synthesized from Oza et al., "Redshifted Sodium Transient"; Oza, discussion, February 3, 2022; Seidel, discussion, February 17, 2022, and March 25, 2022.

second night of Wyttenbach's observations: Wyttenbach et al., "Hot Exoplanet Atmospheres."

With NASA's Nancy Grace Roman: "Roman," NASA, accessed March 27, 2025, https://science.nasa.gov/mission/roman-space-telescope; "Frequently Asked Questions," NASA, accessed January 26, 2026, https://science.nasa.gov/mission/roman-space-telescope/frequently-asked-questions; "About the Roman Space Telescope," NASA, accessed March 27, 2025, https://science.nasa.gov/mission/roman-space-telescope/introducing-the-roman-space-telescope; Penny, discussion.

Some scientists hoped to analyze: Lazzoni, discussion; Lazzoni, email message to author, April 3, 2025; "The Extremely Large Telescope," European Southern Observatory, accessed April 2, 2025, https://elt.eso.org.

another team wanted to examine: The description of the rogue planet research was synthesized from Limbach, discussion; Vanderburg, discussion; Limbach, email message to author, March 18, 2025; Mary Anne Limbach et al., "On the Detection of Exomoons Transiting Isolated Planetary-Mass Objects," *The Astrophysical Journal Letters* 918, no. 2 (September 1, 2021): L25, https://doi.org/10.3847/2041–8213/ac1e2d.

the exomoon search might offer clues: This paragraph was synthesized from J. Laskar, F. Joutel, and P. Robutel, "Stabilization of the Earth's Obliquity by the Moon," *Nature* 361, no. 6413 (February 1993): 615–17, https://doi.org/10.1038/361615a0; J. Laskar and P. Robutel, "The Chaotic Obliquity of the Planets," *Nature* 361, no. 6413 (February 1993): 608–12, https://doi.org/10.1038/361608a0; Keiko Atobe, Shigeru Ida, and Takashi Ito, "Obliquity Variations of Terrestrial Planets in Habitable Zones," *Icarus* 168, no. 2 (April 2004): 223–36, https://doi.org/10.1016/j.icarus.2003.11.017; Miki Nakajima et al., "Large Planets May Not Form Fractionally Large Moons," *Nature Communications* 13, no. 1 (February 1, 2022): 568, https://doi.org/10.1038/s41467-022-28063-8; Teachey, discussion, March 15, 2022; Teachey, email message to author, March 18, 2025.

Kipping's team referred to: Teachey, "Detecting and Characterizing Exomoons"; Kipping et al., "An Exomoon Survey."

a word signifying chaos: Oza, in discussion with the author, June 8, 2022, and January 10, 2023.

In the account that Linda: Morabito, "Discovery of Volcanic Activity," 26.

WASP-49 b's star was close: Oza, discussion, December 15, 2022; Stellarium Web, accessed March 27, 2025, https://stellarium-web.org.

Ecology: The Heroes We Need

Details about the Schell Lab's urban wildlife research were drawn from interviews with team members Chris Schell, Cesar Estien, Lauren Stanton, Gretel Huber, Christine Wilkinson, Tyus Williams, Tali Caspi, Chase Niesner, and Veronica Yovovich, and from my reporting trip to the Schell Lab and its field sites. Details about the coyote dissections and parasite research were drawn from interviews with Yasmine Hentati and my visits to the dissection and parasite labs. General background about urban wildlife and human–wildlife conflict was drawn from interviews with team members and other experts such as Sayantani Basak, Sarah Benson-Amram, Anindita Bhadra, Gabi Fleury, Stan Gehrt, Sarah Jacobson, Seth Riley, and Lewis Stringer. Descriptions of redlining and social justice issues were drawn from interviews with team members and experts such as Jacob Faber, Todd Michney, Thomas Storrs, Maria Miriti, and Rachel Morello-Frosch.

In a black-and-white video: The video was provided by Cesar Estien (PhD student, University of California, Berkeley), email message to author, February 10, 2022.

A skunk approaches a chair: The video was provided by Estien, email, February 10, 2022.

Three raccoons amble: This description was synthesized from Gretel Huber (undergraduate student, University of California, Berkeley) and Lauren Stanton (cognitive ecologist, University of California, Berkeley), in discussion with the author, June 22, 2023; Stanton, discussion, January 31, 2022, and November 29, 2022.

Inside a coyote's small intestine: The description of tapeworm growth was synthesized from Seppo Saari, Anu Näreaho, and Sven Nikander, "Cestoda (Tapeworms)," in *Canine Parasites and Parasitic Diseases* (Elsevier, 2019), 55–81, https://doi.org/10.1016/B978-0-12-814112-0.00004–0; Joann Colville and David Lee Berryhill, *Handbook of Zoonoses: Identification and Prevention* (Mosby, 2007), 182; Małgorzata Samorek-Pieróg et al., "Molecular Confirmation of Massive *Taenia Pisiformis* Cysticercosis in One Rabbit in Poland," *Pathogens* 10, no. 8 (August 14, 2021): 1029, https://doi.org/10.3390/pathogens10081029; Chelsea Wood (University of Washington), email message to author, April 25, 2024; Alessandro Massolo (wildlife ecologist, University of Pisa), in discussion with the author, April 30, 2025; Yasmine Hentati (PhD student, University of Washington), in discussion with the author, January 6, 2023.

who also described himself in his bio: "Christopher Schell," University of California, Berkeley, accessed March 31, 2025, https://ourenvironment.berkeley.edu/users/1757114.

A coyote wandering into a Quiznos: "Coyote Visits Chicago Eatery," *Los Angeles Times*, April 4, 2007, https://www.latimes.com/archives/la-xpm-2007-apr-04-na-coyote4-story.html.

Two raccoons rummaging: Joan Morris, "Raccoons Break into a Redwood City Bank, Leave Empty Handed," *Mercury News*, updated October 21, 2020, https://www.mercurynews.com/2020/10/20/raccoons-break-into-a-redwood-city-bank-leave-empty-handed; Michael Williams, "Raccoons Break into Redwood City Bank, Caught with Paws in Cookie Tin," *San Francisco Chronicle*, October 22, 2020, https://www.sfchronicle.com/bayarea/article/Raccoons-break-into-Redwood-City-bank-caught-15669121.php.

A mountain lion caught: Patrick May, "Another Day, Another Mountain Lion Outside Mark Benioff's SF House," *Mercury News*, November 9, 2017, https://www.mercurynews.com/2017/11/09/another-day-another-mountain-lion-outside-mark-benioffs-sf-house; Dianne de Guzman, "Second San Francisco Mountain Lion Sighting Recorded by Salesforce CEO Marc Benioff," SFGATE, November 8, 2017, https://www.sfgate.com/bayarea/article/san-francisco-mountain-lion-sighting-marc-benioff-12342921.php.

as parts of the globe: Christopher J. Schell et al., "The Evolutionary Consequences of Human–Wildlife Conflict in Cities," *Evolutionary Applications* 14, no. 1 (January 2021): 178–97, https://doi.org/10.1111/eva.13131; Sayantani Basak (wildlife ecologist, Jagiellonian University), in discussion with the author, July 25, 2023.

More than half of humans: "68% of the World Population Projected to Live in Urban Areas by 2050, Says UN," United Nations, accessed April 2, 2025, https://www.un.org/uk/desa/68-world-population-projected-live-urban-areas-2050-says-un.

Expanding cities not only: Sarah Jacobson (animal cognition researcher, City University of New York), in discussion with the author, July 6, 2023; Lewis Stringer (associate director of natural resources, Presidio Trust), in discussion with the author, May 18, 2023; Sarah Benson-Amram (cognitive ecologist, University of British Columbia), in discussion with the author, May 31, 2023.

a problem that disproportionally affected: Jiawen Liu et al., "Disparities in Air Pollution Exposure in the United States by Race/Ethnicity and Income, 1990–2010," *Environmental Health Perspectives* 129, no. 12 (December 2021): 127005, https://doi.org/10.1289/EHP8584.

which could be transmitted to pets: Hentati, discussion, January 6, 2023.

Point Pinole Regional Shoreline: "Overview," Point Pinole Regional Shoreline, accessed April 2, 2025, https://www.ebparks.org/parks/point-pinole.

coyotes have been common: Seth Riley (wildlife ecologist, U.S. National Park Service), in discussion with the author, August 30, 2023; James W. Hody and Roland Kays, "Mapping the Expansion of Coyotes (*Canis Latrans*) across North and Central America," *ZooKeys* 759 (May 22, 2018): 81–97, https://doi.org/10.3897/zookeys.759.15149.

But in other parts: Stan Gehrt (wildlife ecologist, The Ohio State University), in discussion with the author, May 30, 2023.

Then the animals started appearing: This information was synthesized from Gehrt, discussion; Riley, discussion; Hody and Kays, "Mapping the Expansion of Coyotes"; "North American Distribution," Urban Coyote Research Project, accessed April 2, 2025, https://urbancoyoteresearch.com/coyote-info/north-american-distribution; JoAnna Klein, "Coyotes Conquered North America. Now They're Heading South," *New York Times*, May 24, 2018, https://www.nytimes.com/2018/05/24/science/coyotes-americas-spread.html.

They ended up near Navy Pier: "Coyote Captured Near Navy Pier," *Chicago Tribune*, updated August 20, 2021, https://www.chicagotribune.com/2003/12/05/coyote-captured-near-navy-pier; Sarah Zielinski, "Flight Delayed: There's a Coyote on the Runway," *Science News*, April 14, 2015, https://www.sciencenews.org/blog/wild-things/flight-delayed-theres-coyote-runway.

In San Francisco, coyotes: The information about coyotes in San Francisco was synthesized from Stringer, discussion; "Coyotes in the Presidio," Presidio, accessed April 2, 2025, https://presidio.gov/about/sustainability/coyotes-in-the-presidio; Christine E. Wilkinson et al., "Coexistence Across Space and Time: Social-ecological Patterns Within a Decade of Human-Coyote Interactions in San Francisco," *People and Nature* 5, no. 6 (December 2023): 2158–77, https://doi.org/10.1002/pan3.10549; Bianca Taylor, "San Francisco's Coyotes Are Back, and They Are Thriving," KQED, February 20, 2020, https://www.kqed.org/news/11799871/bay-curious-coyotes.

Flourishing animal populations: Gehrt, discussion; Basak, discussion; Schell et al., "The Evolutionary Consequences."

Hunter-gatherers had to ward off: Ann Margvelashvili et al., "An Ancient Cranium from Dmanisi: Evidence for Interpersonal Violence, Disease, and Possible Predation by Carnivores on Early Pleistocene Homo," *Journal of Human Evolution* 166 (May 2022): 103180, https://doi.org/10.1016/j.jhevol.2022.103180; Ann Margvelashvili (senior researcher, Georgian National Museum), email message to author, April 28, 2024; C. K. Brain, "An Attempt to Reconstruct the Behaviour of Australopithecines: The Evidence for Interpersonal Violence," *Zoologica Africana* 7, no. 1 (January 1972): 379–401, https://doi.org/10.1080/00445096.1972.11447451.

Elephants raid crops in Asia: L. Jen Shaffer et al., "Human–Elephant Conflict: A Review of Current Management Strategies and Future Directions," *Frontiers in Ecology and Evolution* 6 (January 11, 2019): 235, https://doi.org/10.3389/fevo.2018.00235; Joshua M. Plotnik and Sarah L. Jacobson, "A 'Thinking Animal' in Conflict: Studying Wild Elephant Cognition in the Shadow of Anthropogenic Change," *Current Opinion in Behavioral Sciences* 46 (August 2022): 101148, https://doi.org/10.1016/j.cobeha.2022.101148; Jacobson, discussion.

In cities, conflicts could encompass: Sayantani M. Basak et al., "Public Perceptions and Attitudes Toward Urban Wildlife Encounters—A Decade of Change," *Science of The Total Environment* 834 (August 2022): 155603, https://doi.org/10.1016/j.scitotenv.2022.155603; Basak, discussion; Calum X. Cunningham et al., "Permanent Daylight Saving Time Would Reduce Deer-Vehicle Collisions," *Current Biology* 32, no. 22 (November 2022): 4982–88.e4, https://doi.org/10.1016/j.cub.2022.10.007.

Coyotes in North America: Connor A. Thompson, Jay R. Malcolm, and Brent R. Patterson, "Individual and Temporal Variation in Use of Residential Areas by Urban Coyotes," *Frontiers in Ecology and Evolution* 9 (June 1, 2021): 687504, https://doi.org/10.3389/fevo.2021.687504; Stewart W. Breck et al., "The Intrepid Urban Coyote: A Comparison of Bold and Exploratory Behavior in Coyotes from Urban and Rural Environments," *Scientific Reports* 9, no. 1 (February 14, 2019): 2104, https://doi.org/10.1038/s41598-019-38543-5; Stringer, discussion; Riley, discussion.

Attacks on people are rare: Stanley D. Gehrt et al., "Severe Environmental Conditions Create Severe Conflicts: A Novel Ecological Pathway to Extreme Coyote Attacks on Humans," *Journal of Applied Ecology* 60, no. 2 (February 2023): 353–64, https://doi.org/10.1111/1365–2664.14333; Jonathan Farr et al., "A Ten-Year Community Reporting Database Reveals Rising Coyote Boldness and Associated Human Concern in Edmonton, Canada," *Ecology and Society* 28, no. 2 (2023): art19, https://doi.org/10.5751/ES-14015–280219; Giulia Bombieri et al., "A Worldwide Perspective on Large Carnivore Attacks on Humans," ed. Andy P. Dobson, *PLOS Biology* 21, no. 1 (January 31, 2023): e3001946, https://doi.org/10.1371/journal.pbio.3001946; Tali Caspi (PhD student, University of California, Davis), in discussion with the author, March 29, 2022. One of the aggressive coyotes is described in Lauren Hernández, "Coyote Killed After Charging Toddlers in S.F.'s Botanical Garden," *San Francisco Chronicle,* updated July 16, 2021, https://www.sfchronicle.com/bayarea/article/Coyote-killed-after-charging-toddlers-in-S-F-s-16320746.php; Rachel Swan, "Carl Was Shot Dead. The Killing Could Lead to a Coyote Population Boom in San Francisco," *San Francisco Chronicle,* updated October 4, 2022, https://www.sfchronicle.com/sf/article/sf-carl-coyote-17484087.php; Wilkinson et al., "Coexistence Across Space." Another coyote is described in Nathan Solis, "Coyote That Attacked Girl at San Francisco's Botanical Garden Is Killed," *Los Angeles Times,* July 2, 2024, https://www.latimes.com/california/story/2024-07-02/san-francisco-coyote-killed-girl-attack; Amy Graff, "One of Three Coyotes Killed in SF Identified as Animal That Attacked Child," *SFGATE,* updated July 2, 2024, https://www.sfgate.com/bayarea/article/coyotes-killed-after-one-bites-child-sf-19549772.php. A third coyote is described in Heather Knight and Loren Elliott, "The Coyotes of San Francisco," *New York Times,* updated May 20, 2025, https://www.nytimes.com/2025/05/19/us/coyotes-san-francisco-california.html; Amy Graff, "Coyote Behind Spate of Attacks in San Francisco Killed by Federal Officials," *SFGATE,* October 10, 2024, https://www.sfgate.com/bayarea/article/san-francisco-crissy-field-coyote-killed-19827801.php.

Instead of killing or moving: The paragraph about coexistence was synthesized from Basak, discussion; Benson-Amram, discussion; Stringer, discussion; Christine Wilkinson (postdoctoral researcher, University of California, Santa Cruz), in discussion with the author, January 15, 2022; Chris Schell (urban ecologist, University of California, Berkeley), in discussion with the author, January 19, 2022.

Coyotes were common: Washington Department of Fish and Wildlife, "Tips for Living with Coyotes," *Medium*, January 27, 2023, https://wdfw.medium.com/tips-for-living-with-coyotes-e99a20f2ae89; "Predators," Bridle Trails Park Foundation, accessed April 2, 2025, https://www.bridletrails.org/naturalist-predators.

the exact line, spoken: "The Dark Knight—'Gotham Needs Its True Hero.' (480p)," Franchise Fan, July 18, 2012, YouTube video, 4:13, https://www.youtube.com/watch?v=08uyGm0BRp4.

On an online message board: "What Was the Meaning of This Quote from 'the Dark Knight' Movie: 'Because He's the Hero Gotham Needs, but Not the One It Deserves Right Now'?," *Quora*, accessed April 2, 2025, https://www.quora.com/What-was-the-meaning-of-this-quote-from-%E2%80%9CThe-Dark-Knight%E2%80%9D-movie-Because-hes-the-hero-Gotham-needs-but-not-the-one-it-deserves-right-now; "Shouldn't Gordon Have Said 'He's the Hero Gotham NEEDS, But Not the One It DESERVES Right Now' Instead of 'He's the Hero Gotham Deserves, But Not the One It Needs Right Now' for the Batman in the Dark Knight?," *Quora*, accessed April 2, 2025, https://www.quora.com/Shouldnt-Gordon-have-said-Hes-the-hero-Gotham-NEEDS-but-not-the-one-it-DESERVES-right-now-instead-of-Hes-the-hero-Gotham-deserves-but-not-the-one-it-needs-right-now-for-the-Batman-in-The-Dark-Knight.

Heavy metals such as lead: Hentati, discussion, January 6, 2023.

some parasites could be transmitted: Hentati, discussion, January 6, 2023.

Coyotes in parts of: Alessandro Massolo et al., "European *Echinococcus multilocularis* Identified in Patients in Canada," *New England Journal of Medicine* 381, no. 4 (July 25, 2019): 384–85, https://doi.org/10.1056/NEJMc1814975; Maria A. Santa et al., "It's a Small World for Parasites: Evidence Supporting the North American Invasion of European *Echinococcus multilocularis*," *Proceedings of the Royal Society B: Biological Sciences* 290, no. 1994 (March 8, 2023): 20230128, https://doi.org/10.1098/rspb.2023.0128; Massolo, discussion; "Echinococcosis," World Health Organization, accessed May 1, 2025, https://www.who.int/news-room/fact-sheets/detail/echinococcosis; "Tapeworm in Coyotes That Can Cause Fatal Tumours in People 'Has Spread All over Alberta,'" *CBC News*, updated July 25, 2019, https://www.cbc.ca/news/canada/calgary/tapeworm-echinococcus-multilocularis-alberta-klein-calgary

-veterinarian-disease-coyotes-dogs-1.5224864. For cases found in dogs, also see Michelle D. Evason et al., "Emerging *Echinococcus* Tapeworms: Fecal PCR Detection of *Echinococcus multilocularis* in 26 Dogs from the United States and Canada (2022–2024)," *Journal of the American Veterinary Medical Association* 263, no. 2 (February 1, 2025): 1–5, https://doi.org/10.2460/javma.24.07.0471; Temitope U. Kolapo et al., "Canine Alveolar Echinococcosis: An Emerging and Costly Introduced Problem in North America," ed. Walter Lilenbaum, *Transboundary and Emerging Diseases* 2023 (February 21, 2023): 1–10, https://doi.org/10.1155/2023/5224160.

Roundworms transmitted by raccoons: Emily L. Blizzard et al., "Geographic Expansion of *Baylisascaris procyonis* Roundworms, Florida, USA," *Emerging Infectious Diseases* 16, no. 11 (November 2010): 1803–4, https://doi.org/10.3201/eid1611.100549; Frank Sorvillo, "*Baylisascaris procyonis*: An Emerging Helminthic Zoonosis," *Emerging Infectious Diseases* 8, no. 4 (April 2002): 355–59, https://doi.org/10.3201/eid0804.010273; "How Raccoon Roundworm Spreads," U.S. Centers for Disease Control and Prevention, November 5, 2024, https://www.cdc.gov/baylisascaris/causes.

In 2020, Chris and seven: Christopher J. Schell et al., "The Ecological and Evolutionary Consequences of Systemic Racism in Urban Environments," *Science* 369, no. 6510 (September 18, 2020): eaay4497, https://doi.org/10.1126/science.aay4497.

a discriminatory policy called redlining: "Mapping Inequality: Redlining in New Deal America," *Mapping Inequality*, accessed April 2, 2025, https://dsl.richmond.edu/panorama/redlining; Todd M. Michney, "How the City Survey's Redlining Maps Were Made: A Closer Look at HOLC's Mortgagee Rehabilitation Division," *Journal of Planning History* 21, no. 4 (November 2022): 316–44, https://doi.org/10.1177/15385132211013361; Todd M. Michney, "How and Why the Home Owners' Loan Corporation Made Its Redlining Maps," *Mapping Inequality*, accessed April 2, 2025, https://dsl.richmond.edu/panorama/redlining/howandwhy; Schell, "Ecological and Evolutionary Consequences."

The best-preserved records of redlining: Details about HOLC's grading system were synthesized from Robert K. Nelson, "Introduction," *Mapping Inequality*, accessed April 2, 2025, https://dsl.richmond.edu/panorama/redlining/introduction; Michney, "How the City Survey's"; Michney, "How and Why"; Jacob W. Faber, "We Built This: Consequences of New Deal Era Intervention in America's Racial Geography," *American Sociological Review* 85, no. 5 (October 2020): 739–75, https://doi.org/10.1177/0003122420948464; Louis Lee Woods, "The Federal Home Loan Bank Board, Redlining, and the National Proliferation of Racial Lending Discrimination, 1921–1950,"

Journal of Urban History 38, no. 6 (November 2012): 1036–59, https://doi.org/10.1177/0096144211435126; Todd Michney (historian, Georgia Tech), in discussion with the author, May 2, 2024; Thomas Storrs (historian, University of Virginia), in discussion with the author, April 22, 2024; Jacob Faber (sociologist, New York University), in discussion with the author, January 7, 2025.

HOLC's written reports about Seattle: "Seattle," *Mapping Inequality*, accessed April 2, 2025, https://dsl.richmond.edu/panorama/redlining/map/WA/Seattle/area_descriptions/#loc=11/47.5936/-122.3411. Specific quotes can be found by typing them into the "Search the area descriptions" field.

in the San Francisco Bay Area: "Oakland," *Mapping Inequality*, accessed April 2, 2025, https://dsl.richmond.edu/panorama/redlining/map/CA/Oakland/area_descriptions/#loc=11/37.8099/-122.2263; "San Francisco," *Mapping Inequality*, accessed April 2, 2025, https://dsl.richmond.edu/panorama/redlining/map/CA/SanFrancisco/area_descriptions#loc=12/37.7584/-122.4368. Specific quotes can be found by typing them into the "Search the area descriptions" field.

Even "good" neighborhoods: Faber, discussion.

One San Francisco area: "San Francisco: Area C4," *Mapping Inequality*, accessed April 2, 2025, https://dsl.richmond.edu/panorama/redlining/map/CA/SanFrancisco/area_descriptions/C4#mapview=full&loc=12/37.7584/-122.437&scan=3/64.4281/-120.

HOLC wanted a methodical way: Michney, "How the City Survey's"; Woods, "The Federal Home Loan Bank Board"; Faber, "We Built This." Note that HOLC didn't use the ratings to systematically deny refinancing to "C" and "D" neighborhoods, since the maps were created after nearly all the loans had been processed; see Michney, "How the City Survey's"; Michney, "How and Why."

HOLC's own data showed: Michney, email message to fact-checker, March 11, 2025; Michney, "How and Why."

The Federal Housing Administration: Price Fishback et al., "New Evidence on Redlining by Federal Housing Programs in the 1930s," *Journal of Urban Economics* 141 (May 2024): 103462, https://doi.org/10.1016/j.jue.2022.103462; Amy E. Hillier, "Redlining and the Home Owners' Loan Corporation," *Journal of Urban History* 29, no. 4 (May 2003): 394–420, https://doi.org/10.1177/0096144203029004002; Michney, "How and Why"; Michney, discussion; Storrs, discussion.

Government officials encouraged: Michney, "How and Why"; Michney, discussion; Woods, "The Federal Home Loan Bank Board."

While Black buyers: Faber, "We Built This"; Faber, discussion; Storrs, discussion; Michney, discussion. Note that scholars are still debating how much the physical HOLC maps themselves—the most infamous examples of redlining, largely because the documents have survived and been digitized—drove lending decisions and inequality in cities. The documents were not supposed to be widely shared with private industry, but some researchers argue that the agency's maps still could have exerted influence in various ways. What experts generally agree on is that HOLC and other federal officials spread and endorsed the racist ideas behind redlining. At a minimum, the maps can be seen as a partial proxy for the widespread practice of redlining in private industry and at government agencies. For a summary of the debate, see the background sections of Daniel Aaronson, Daniel Hartley, and Bhashkar Mazumder, "The Effects of the 1930s HOLC 'Redlining' Maps," *American Economic Journal: Economic Policy* 13, no. 4 (November 1, 2021): 355–92, https://doi.org/10.1257/pol.20190414; and Daniel Aaronson et al., "The Long-Run Effects of the 1930s Redlining Maps on Children," *Journal of Economic Literature* 61, no. 3 (September 1, 2023): 846–62, https://doi.org/10.1257/jel.20221702. Relevant studies include Kenneth T. Jackson, "Race, Ethnicity, and Real Estate Appraisal: The Home Owners Loan Corporation and the Federal Housing Administration," *Journal of Urban History* 6, no. 4 (August 1980): 419–52, https://doi.org/10.1177/009614428000600404; Hillier, "Redlining and the Home Owners'"; Jennifer S. Light, "Nationality and Neighborhood Risk at the Origins of FHA Underwriting," *Journal of Urban History* 36, no. 5 (September 2010): 634–71, https://doi.org/10.1177/0096144210365677; James Greer, "The Home Owners' Loan Corporation and the Development of the Residential Security Maps," *Journal of Urban History* 39, no. 2 (March 2013): 275–96, https://doi.org/10.1177/0096144212436724; Daniel Aaronson et al., "The Long-Run Effects of the 1930s HOLC 'Redlining' Maps on Place-Based Measures of Economic Opportunity and Socioeconomic Success," *Regional Science and Urban Economics* 86 (January 2021): 103622, https://doi.org/10.1016/j.regsciurbeco.2020.103622; Disa M Hynsjö and Luca Perdoni, "The Effects of Federal 'Redlining' Maps: A New Empirical Strategy," American Economic Association, accessed April 16, 2025, https://www.aeaweb.org/conference/2023/program/2084; Faber, "We Built This"; Michney, "How the City Survey's"; Michney, "How and Why"; Woods, "The Federal Home Loan Bank Board"; and Fishback et al., "New Evidence on Redlining." My understanding of the debate was also informed by Michney, discussion; Storrs, discussion; Faber, discussion; "Mapping Inequality: Redlining in New Deal America."

A white family: Michney, discussion; Faber, discussion.

A Black family: Michney, discussion; Faber, discussion; Storrs, discussion;

Junia Howell and Elizabeth Korver-Glenn, "The Increasing Effect of Neighborhood Racial Composition on Housing Values, 1980–2015," *Social Problems* 68, no. 4 (October 19, 2021): 1051–71, https://doi.org/10.1093/socpro/spaa033.

The wealth gap: Faber, discussion; Junia Howell, "Color Coded: The Growing Racial Inequality in Home Appraisals," Consumer Financial Protection Bureau, January 24, 2023, https://files.consumerfinance.gov/f/documents/cfpb_appraisal-hearing_junia-howell-testimony_2023-01-24.pdf.

Chris's team found that: Schell et al., "Ecological and Evolutionary Consequences."

other researchers had reported links: Rachel Morello-Frosch (environmental health scientist, University of California, Berkeley), in discussion with the author, July 28, 2023; Anthony Nardone et al., "Associations Between Historical Residential Redlining and Current Age-Adjusted Rates of Emergency Department Visits Due to Asthma Across Eight Cities in California: An Ecological Study," *The Lancet Planetary Health* 4, no. 1 (January 2020): e24–31, https://doi.org/10.1016/S2542–5196(19)30241–4.

Those animals could, in turn: Schell et al., "Ecological and Evolutionary Consequences."

Cesar examined data from iNaturalist: Cesar O. Estien et al., "Historical Redlining Is Associated with Disparities in Wildlife Biodiversity in Four California Cities," *Proceedings of the National Academy of Sciences* 121, no. 25 (June 18, 2024): e2321441121, https://doi.org/10.1073/pnas.2321441121.

Previous research had suggested: Jeremy S. Hoffman, Vivek Shandas, and Nicholas Pendleton, "The Effects of Historical Housing Policies on Resident Exposure to Intra-Urban Heat: A Study of 108 US Urban Areas," *Climate* 8, no. 1 (January 13, 2020): 12, https://doi.org/10.3390/cli8010012; Haley M. Lane et al., "Historical Redlining Is Associated with Present-Day Air Pollution Disparities in U.S. Cities," *Environmental Science & Technology Letters* 9, no. 4 (April 12, 2022): 345–50, https://doi.org/10.1021/acs.estlett.1c01012; Anthony Nardone et al., "Redlines and Greenspace: The Relationship Between Historical Redlining and 2010 Greenspace Across the United States," *Environmental Health Perspectives* 129, no. 1 (January 2021): 017006, https://doi.org/10.1289/EHP7495.

In another 2020 paper: Christopher J. Schell et al., "Recreating Wakanda by Promoting Black Excellence in Ecology and Evolution," *Nature Ecology & Evolution* 4, no. 10 (July 24, 2020): 1285, https://doi.org/10.1038/s41559-020-1266-7.

Coyotes have flourished in cities: The information in this paragraph was synthesized from Gehrt, discussion; Riley, discussion; Stringer, discussion; Gehrt et al., "Severe Environmental Conditions."

One question occupying biologists: Stanton, discussion, January 31, 2022; Benson-Amram, discussion.

given that domestic dogs: Data on the frequency of coyote attacks can be found in Rex O. Baker and Robert M. Timm, "Coyote Attacks on Humans, 1970–2015," *Proceedings of the Vertebrate Pest Conference* 27 (2016), https://doi.org/10.5070/V427110675; Bombieri et al., "A Worldwide Perspective." Data on the frequency of domestic dog bites can be found in Peter S. Tuckel and William Milczarski, "The Changing Epidemiology of Dog Bite Injuries in the United States, 2005–2018," *Injury Epidemiology* 7, no. 1 (December 2020): 57, https://doi.org/10.1186/s40621-020-00281-y; J. Gilchrist, J. J. Sacks, D. White, and M.-J. Kresnow, "Dog Bites: Still a Problem?," *Injury Prevention* 14 (2008): 296–301.

In Los Angeles, an anti-coyote: Chase Alexander Niesner, "At Home with Coyotes: An Exploration of Human-Coyote Relations in the Los Angeles 'Ecology of Selves'" (PhD diss., University of California, Los Angeles, 2023), 52–53, https://escholarship.org/uc/item/74c8g1c0; Louis Sahagún, "Inside the War Against Southern California's Urban Coyotes. 'Horrific' or Misunderstood?," *Los Angeles Times*, September 20, 2022, https://www.latimes.com/environment/story/2022-09-20/southern-california-coyotes-population-escalating-war.

in one of Chris's papers: Schell et al., "Ecological and Evolutionary Consequences"; Meredith C. VanAcker et al., "Enhancement of Risk for Lyme Disease by Landscape Connectivity, New York, New York, USA," *Emerging Infectious Diseases* 25, no. 6 (June 2019): 1136–43, https://doi.org/10.3201/eid2506.181741.

Chris liked to quote: "'Wealth, Race, and Wildlife: The Impacts of Structural Inequality on Urban Wildlife'—Chris Schell," NCEAS, March 3, 2021, YouTube video, 1:03:36, https://www.youtube.com/watch?v=SAxBNfzBq7A&t=2936s.

characters had largely separate storylines: Maya Phillips, "The Narrative Experiment That Is the Marvel Cinematic Universe," *New Yorker*, April 26, 2019, https://www.newyorker.com/culture/culture-desk/the-narrative-experiment-that-is-the-marvel-cinematic-universe.

"The actions of each": Sean Howe, *Marvel Comics: The Untold Story* (Harper Perennial, 2012), 47.

as one critic described it: Phillips, "The Narrative Experiment."

Climate Change: Emerald Dreams

Details about the Emerald Tutu project and the Oregon State University wave lab were drawn from interviews with team members Gabriel Cira, Julia Hopkins, Jena Tegeler, Nick Lutsko, and Louiza Wise and lab staff members Pedro Lomonaco and Tim Maddux; from my reporting trips to the wave lab and the Boston area; from the team's website, https://emerald-tutu.com; and from the team's scientific paper Julia Hopkins et al., "The Emerald Tutu: Floating Vegetated Canopies for Coastal Wave Attenuation," *Frontiers in Built Environment* 8 (June 9, 2022): 885298, https://doi.org/10.3389/fbuil.2022.885298. Biographical details about team members were drawn from interviews with those people. Descriptions of coastal issues, protection, science, and policy were drawn from interviews with those team members or experts such as Dan Cox, Tori Tomiczek, Kiernan Kelty, Edoardo Borgomeo, Stephanie Kruel, Greg Guannel, Siddharth Narayan, Katie Arkema, Ariana Sutton-Grier, Evelien Brand, and Renske de Winter.

a competition at MIT: Climate Changed, archived April 4, 2023, at the *Wayback Machine*, https://web.archive.org/web/20230404223339/http://climatechangedmit.com.

an architecture firm in Paris: "R&Sie(n)," San Francisco Museum of Modern Art, accessed August 15, 2024, https://www.sfmoma.org/artist/R_Sie_n; "Profile: François Roche and R&Sie(n)," *Icon*, August 8, 2011, https://www.iconeye.com/architecture/profile-francois-roche-and-r-sie-n.

house enveloped by hydroponic ferns: Catherine Slessor, "'I'm Lost in Paris' House by R&Sie(n), Paris, France," October 1, 2009, https://www.architectural-review.com/today/im-lost-in-paris-house-by-rsien-paris-france; "I'm Lost in Paris / R&Sie(n)," *ArchDaily*, January 23, 2009, https://www.archdaily.com/12212/im-lost-in-paris-rsien.

Hurricane Bob had blasted: Eric Williams, "'I Got Bobbed!': Tales of the Hurricane That Steamrolled Cape Cod 30 Years Ago," *Cape Cod Times*, August 16, 2021, https://www.capecodtimes.com/story/news/2021/08/16/recollections-hurricane-bob-steamrolled-cape-cod-30-years-1991-wind-erosion-damage-wellfleet/5552292001; I. Valiela et al., "Ecological Effects of Major Storms on Coastal Watersheds and Coastal Waters: Hurricane Bob on Cape Cod," *Journal of Coastal Research* 14, no. 1 (1998): 218–38, http://www.jstor.org/stable/4298771.

cities had typically staved off: Hopkins et al., "The Emerald Tutu"; Ariana E. Sutton-Grier, Kateryna Wowk, and Holly Bamford, "Future of Our Coasts: The Potential for Natural and Hybrid Infrastructure to Enhance the Resilience of Our Coastal Communities, Economies, and Ecosystems," *Environmental Science & Policy* 51 (August 2015): 137–48, https://doi.org/10.1016/j.envsci.2015.04.006.

They were built to withstand: Julia Hopkins (coastal engineer, Northeastern University), in discussion with the author, June 8, 2021.

If forecasts were off: Hopkins, discussion.

In recent years, researchers had: The information in this paragraph was synthesized from Sutton-Grier, Wowk, and Bamford, "Future of Our Coasts"; Ariana Sutton-Grier et al., "Investing in Natural and Nature-Based Infrastructure: Building Better Along Our Coasts," *Sustainability* 10, no. 2 (February 15, 2018): 523, https://doi.org/10.3390/su10020523; Cindy M. Palinkas et al., "Innovations in Coastline Management with Natural and Nature-Based Features (NNBF): Lessons Learned from Three Case Studies," *Frontiers in Built Environment* 8 (April 27, 2022): 814180, https://doi.org/10.3389/fbuil.2022.814180; Siddharth Narayan (coastal engineer, East Carolina University), in discussion with the author, November 29, 2021, and September 18, 2023; Katie Arkema (Earth scientist, Pacific Northwest National Laboratory), in discussion with the author, November 30, 2021.

But some urban shorelines: Gabriel Cira (independent architect), in discussion with the author, September 6, 2023.

Emerald Necklace, a set of parks: "Emerald Necklace," City of Boston, updated July 14, 2016, https://www.boston.gov/environment-and-energy/emerald-necklace.

a video of a drag queen: "The Emerald Tutu," ARCH CIRA, April 25, 2018, Vimeo video, 3:26, https://vimeo.com/266553400.

To counteract dune erosion: Evelien Brand, Gemma Ramaekers, and Quirijn Lodder, "Dutch Experience with Sand Nourishments for Dynamic Coastline Conservation—An Operational Overview," *Ocean & Coastal Management* 217 (February 2022): 106008, https://doi.org/10.1016/j.ocecoaman.2021.106008.

these algae often anchored: "Algae vs Seagrass," National Museum of Natural History, accessed August 15, 2024, https://ocean.si.edu/holding-tank/images-hide/algae-vs-seagrass; Timothy Baxter, Martin Coombes, and Heather Viles, "Managing Marine Growth on Historic Maritime Structures: An Assessment of Perceptions and Current Management Practices," *Frontiers in Marine Science* 9 (June 6, 2022): 913972, https://doi.org/10.3389/fmars.2022.913972.

But by fixing a problem: The description of potential problems with hard infrastructure was synthesized from Greg Guannel (coastal engineer, University of the Virgin Islands), in discussion with the author, November 29, 2021; Hopkins, in discussion with the author, November 10, 2021; Ariana Sutton-Grier (ecologist, University of Maryland), in discussion with the author, December 1, 2021; Arkema, discussion.

The Netherlands—which is: Kimmelman, "Dutch Have Solutions"; Renske de Winter (coastal engineer, Deltares), in discussion with the author, July 26, 2022.

Nature-based solutions: The description of the benefits of nature-based solutions was synthesized from Sutton-Grier, Wowk, and Bamford, "Future of Our Coasts"; Narayan, discussion, November 29, 2021, and September 18, 2023; Arkema, discussion.

More than sixty billion dollars: Benjamin H. Strauss et al., "Economic Damages from Hurricane Sandy Attributable to Sea Level Rise Caused by Anthropogenic Climate Change," *Nature Communications* 12, no. 1 (May 18, 2021): 2720, https://doi.org/10.1038/s41467-021-22838-1.

Trillions of dollars: Michelle A. Hummel et al., "Economic Evaluation of Sea-Level Rise Adaptation Strongly Influenced by Hydrodynamic Feedbacks," *Proceedings of the National Academy of Sciences* 118, no. 29 (July 20, 2021): e2025961118, https://doi.org/10.1073/pnas.2025961118.

The flooding risks: The information in this paragraph was synthesized from Hopkins, discussion, November 10, 2021; Nick Lutsko (climate scientist, Scripps Institution of Oceanography), in discussion with the author, November 22, 2021; Narayan, discussion, November 29, 2021, and September 28, 2023; Sutton-Grier, discussion; Leonard O. Ohenhen et al., "Hidden Vulnerability of US Atlantic Coast to Sea-Level Rise Due to Vertical Land Motion," *Nature Communications* 14, no. 1 (April 11, 2023): 2038, https://doi.org/10.1038/s41467-023-37853-7.

a report on Boston's climate: "Climate Ready Boston," City of Boston, December 2016, xxix, https://www.boston.gov/sites/default/files/file/2023/03/2016_climate_ready_boston_report.pdf.

In the South Boston neighborhood: "Climate Ready Boston," 285; Hannah Wagner (climate resilience project manager, City of Boston Environment Department), in discussion with the author, October 4, 2023.

Electricity and public transportation: Sutton-Grier, discussion.

Hurricane Sandy had temporarily dragged: "Economic Impact of Hurricane Sandy," U.S. Department of Commerce, September 2013, https://www.commerce.gov/sites/default/files/migrated/reports/sandyfinal101713.pdf; Christopher Matthews, "Better Safe than Sorry: Why U.S. Securities Exchanges Closed for Second—and Maybe Third—Straight Day," *Time*, October 30, 2012, https://business.time.com/2012/10/30/better-safe-than-sorry-u-s-securities-exchanges-close-for-second-straight-day; Sutton-Grier, discussion.

Flooded roads could: Arkema, discussion.

After Hurricane Ida: Nicholas Bogel-Burroughs and Katy Reckdahl, "The Greatest Killer in New Orleans Wasn't the Hurricane. It Was the Heat," *New York Times*, September 15, 2021, https://www.nytimes.com/2021/09/15/us/new-orleans-hurricane-ida-heat.html.

wetlands prevented more than: Siddharth Narayan et al., "The Value of Coastal Wetlands for Flood Damage Reduction in the Northeastern USA," *Scientific Reports* 7, no. 1 (August 31, 2017): 9463, https://doi.org/10.1038/s41598-017-09269-z.

$934 million: Zaid Al-Attabi et al., "The Impacts of Tidal Wetland Loss and Coastal Development on Storm Surge Damages to People and Property: A Hurricane Ike Case-Study," *Scientific Reports* 13, no. 1 (March 21, 2023): 4620, https://doi.org/10.1038/s41598-023-31409-x.

In one study of Florida Keys: Tori Tomiczek et al., "Rapid Damage Assessments of Shorelines and Structures in the Florida Keys after Hurricane Irma," *Natural Hazards Review* 21, no. 1 (February 2020): 05019006, https://doi.org/10.1061/(ASCE)NH.1527–6996.0000349.

Those plant-protected houses: Tomiczek et al., "Rapid Damage Assessments."

One team analyzed sixty-nine studies: Siddharth Narayan et al., "The Effectiveness, Costs, and Coastal Protection Benefits of Natural and Nature-Based Defences," *PLOS One* 11, no. 5 (May 2, 2016): e0154735, https://doi.org/10.1371/journal.pone.0154735.

one recent lab study: Bregje K. Van Wesenbeeck et al., "Wave Attenuation Through Forests Under Extreme Conditions," *Scientific Reports* 12, no. 1 (February 3, 2022): 1884, https://doi.org/10.1038/s41598-022-05753-3.

Performance might decline: Narayan, discussion, November 29, 2021.

Natural systems were limited: Narayan, discussion, November 29, 2021.

At certain sites: Narayan, discussion, September 18, 2023; Al-Attabi et al., "The Impacts of Tidal Wetland Loss."

they might not be able: Stephanie Kruel (senior regulatory and resilience advisor, VHB), in discussion with the author, November 18, 2021.

If sea levels rose too quickly: Sarah E. Hobbie and Nancy B. Grimm, "Nature-Based Approaches to Managing Climate Change Impacts in Cities," *Philosophical Transactions of the Royal Society B: Biological Sciences* 375, no. 1794 (March 16, 2020): 20190124, https://doi.org/10.1098/rstb.2019.0124.

Based on the data: The results described in this paragraph were synthesized from Hopkins et al., "The Emerald Tutu"; Hopkins, in discussion with the author, July 20, 2022, and September 15, 2023; The Emerald Tutu, email message, November 1, 2021.

The site used to be: "Hess Oil Site: Berm Elevation," City of Boston, February 6, 2019, https://www.boston.gov/sites/default/files/embed/file/2019–02/hess_oil_site_noi_combined.pdf.

An article I'd recently read: Ezra Klein, "It Seems Odd That We Would Just Let the World Burn," *New York Times*, July 15, 2021, https://www.nytimes.com/2021/07/15/opinion/climate-change-energy-infrastructure.html.

Physics: Paper Labyrinth

Details about the team's paper-crumpling research were drawn from interviews or email correspondence with Madelyn Leembruggen, Chris Rycroft, Jovana Andrejevic Kim, Arshad Kudrolli, Amit Dawadi, or Shmuel Rubinstein; my reporting trip to the Clark University lab; sitting in on Zoom meetings with team members; the dissertation Madelyn Jane Leembruggen, "Buckling, Wrinkling, and Crumpling of Simulated Thin Sheets" (PhD diss., Harvard University, 2024), https://dash.lib.harvard.edu/handle/1/37378965; or from the team's scientific papers Omer Gottesman et al., "A State Variable for Crumpled Thin Sheets," *Communications Physics* 1, no. 1 (December 2018): 70, https://doi.org/10.1038/s42005-018-0072-x; Jovana Andrejevic et al., "A Model for the Fragmentation Kinetics of Crumpled Thin Sheets," *Nature Communications* 12, no. 1 (December 2021): 1470, https://doi.org/10.1038/s41467-021-21625-2; Jovana Andrejevic and Chris H. Rycroft, "Simulation of Crumpled Sheets via Alternating Quasistatic and Dynamic Representations," *Journal of Computational Physics* 471 (December 2022): 111607, https://doi.org/10.1016/j.jcp.2022.111607; Madelyn Leembruggen et al., "Computational Model of Twisted Elastic Ribbons," *Physical Review E* 108 (July 24, 2023): 015003, https://doi.org/10.1103/PhysRevE.108.015003. Biographical details about Leembruggen were drawn from interviews with her. General background about physics concepts and crumpling research was drawn from interviews with team members or experts such as Sidney Nagel, Martine Ben Amar, Narayanan Menon, Tom Witten, Anne Meeussen, Andrea Liu, Ian Tobasco, Yoav Lahini, and James Sethna.

their collaborators had crushed: The description of the study was synthesized from Gottesman et al., "A State Variable"; Madelyn Leembruggen (physicist and science communicator), in discussion with the author, March 22, 2021, and September 6, 2023; Jovana Andrejevic Kim (postdoctoral researcher, University of Pennsylvania), in discussion with the author, June 14, 2021; Chris Rycroft (applied mathematician, University of Wisconsin–Madison), in discussion with the author, June 7, 2021.

crumpled paper's everyday nature: Sidney Nagel (physicist, University of Chicago), in discussion with the author, September 26, 2023; Martine Ben Amar (theoretical physicist, École Normale Supérieure), in discussion with the author, September 22, 2023.

coffee rings on tables: Robert D. Deegan et al., "Capillary Flow as the Cause of Ring Stains from Dried Liquid Drops," *Nature* 389, no. 6653 (October 23, 1997): 827–29, https://doi.org/10.1038/39827.

while paper crumpling might seem: Rycroft, discussion; Anne Meeussen (physicist, Harvard University), in discussion with the author, October 20, 2023.

If we could grasp: The information in this paragraph was synthesized from Nagel, discussion; Meeussen, discussion; Narayanan Menon (physicist, University of Massachusetts Amherst), in discussion with the author, October 4, 2023; Ian Tobasco (applied mathematician, Rutgers University), in discussion with the author, October 2, 2023.

at the microscopic level, graphene: The information about graphene was synthesized from Rycroft, discussion; Nagel, discussion; Jianfeng Zang et al., "Multifunctionality and Control of the Crumpling and Unfolding of Large-Area Graphene," *Nature Materials* 12, no. 4 (April 2013): 321–25, https://doi.org/10.1038/nmat3542.

a tongue-in-cheek paper: Madelyn Leembruggen and Caroline Martin, "What's for Lunch? A Systematic Ordering of Foods in the Soup-Salad-Sandwich Phase Space," arXiv (March 30, 2022), http://arxiv.org/abs/2203.16580.

theoretical particles called axions: The information about axions was synthesized from Leembruggen, discussion, May 21, 2021; Joshua Eby et al., "Collapse of Axion Stars," *Journal of High Energy Physics* 2016, no. 12 (December 2016): 66, https://doi.org/10.1007/JHEP12(2016)066; Francesca Chadha-Day, John Ellis, and David J. E. Marsh, "Axion Dark Matter: What Is It and Why Now?," *Science Advances* 8, no. 8 (February 25, 2022): eabj3618, https://doi.org/10.1126/sciadv.abj3618.

Her result contradicted: Pierre-Henri Chavanis, "Collapse of a Self-Gravitating Bose-Einstein Condensate with Attractive Self-Interaction," *Physical Review D* 94, no. 8 (October 20, 2016): 083007, https://doi.org/10.1103/PhysRevD.94.083007; Eby et al., "Collapse of Axion Stars."

little fluctuations in the density: Leembruggen, email messages to author, December 3 and 11, 2024; Nicholas Rapidis (physics PhD student, Stanford University), email messages to author, October 31 and November 5, 2024.

Interest in crumpling: T. A. Witten, "Stress Focusing in Elastic Sheets," *Reviews of Modern Physics* 79, no. 2 (April 25, 2007): 643–75, https://doi.org/10.1103/RevModPhys.79.643.

On the science side: The information in this paragraph was synthesized from Tobasco, discussion; Nagel, discussion; Galileo Galilei, *Dialogues Concerning Two New Sciences* (New York: Macmillan, 1914), 115–16; David E. Alexander, *Nature's Machines: An Introduction to Organismal Biomechanics* (Academic Press, 2017), 42–43; Gwynne Watkins, "'Star Wars' Turns 40: Dirty Secrets from Trash Compactor Scene," *Yahoo! News*, May 25, 2017, https://www.yahoo.com/news/star-wars-turns-40-dirty-secrets-trash-compactor-scene-144820052.html.

But what about thin sheets: Meeussen, discussion.

These questions were part of: The information about complex systems was synthesized from Yoav Lahini (physicist, Tel Aviv University), in discussion with the author, October 5, 2023; Andrea Liu (physicist, University of Pennsylvania), in discussion with the author, September 21, 2023.

In the 1990s: Witten, "Stress Focusing in Elastic Sheets."

Martine Ben Amar (the physicist...): M. Ben Amar and Y. Pomeau, "Crumpled Paper," *Proceedings of the Royal Society A: Mathematical, Physical and Engineering Sciences* 453, no. 1959 (1997): 729–55, https://doi.org/10.1098/rspa.1997.0041; Ben Amar, discussion; Tom Witten (professor emeritus of physics, University of Chicago), in discussion with the author, September 25, 2023.

Witten's team investigated: Witten, discussion; Alex Lobkovsky et al., "Scaling Properties of Stretching Ridges in a Crumpled Elastic Sheet," *Science* 270, no. 5241 (December 1995): 1482–85, https://doi.org/10.1126/science.270.5241.1482.

Lab experiments on sheet crumpling: The information in this paragraph was synthesized from Anne Dominique Cambou and Narayanan Menon, "Three-Dimensional Structure of a Sheet Crumpled into a Ball," *Proceedings of the National Academy of Sciences* 108, no. 36 (September 6, 2011): 14741–45, https://doi.org/10.1073/pnas.1019192108; Menon, discussion.

Nagel's team had crumpled: The information in this paragraph was synthesized from Kittiwit Matan et al., "Crumpling a Thin Sheet," *Physical Review Letters* 88, no. 7 (January 30, 2002): 076101, https://doi.org/10.1103/PhysRevLett.88.076101; Nagel, discussion; Lahini, discussion.

physicists Shmuel Rubinstein: The information about the piston study over the next few paragraphs was synthesized from Gottesman et al., "A State Variable"; Andrejevic et al., "Model for the Fragmentation Kinetics"; Leembruggen, discussion, September 6, 2023; Rycroft, discussion; Andrejevic Kim, discussion; Shmuel Rubinstein (physicist, Hebrew University of Jerusalem), email messages to author, December 4 and 7, 2024.

the Warren-girder case: Leembruggen, "Buckling, Wrinkling, and Crumpling," 120.

Jovana had confirmed: Andrejevic Kim, discussion; Gottesman et al., "A State Variable."

Originally from Serbia: Adam Zewe, "Grad Student Profile: Jovana Andrejevic," Harvard University, July 2, 2018, https://seas.harvard.edu/news/2018/07/grad-student-profile-jovana-andrejevic.

Jovana had started investigating: The information about Jovana's facet research over the next several paragraphs was synthesized from Andrejevic Kim, discussion; Rycroft, discussion; Leembruggen, discussion, March 22, 2021;

Andrejevic et al., "Model for the Fragmentation Kinetics"; Andrejevic and Rycroft, "Simulation of Crumpled Sheets"; E. Kaminski and C. Jaupart, "The Size Distribution of Pyroclasts and the Fragmentation Sequence in Explosive Volcanic Eruptions," *Journal of Geophysical Research: Solid Earth* 103, no. B12 (December 10, 1998): 29759–79, https://doi.org/10.1029/98JB02795; Liu, discussion.

Marcel Duchamp, the French artist: Terry Riggs, "Marcel Duchamp," Tate Gallery, October 1997, https://www.tate.org.uk/art/artists/marcel-duchamp-1036.

quasicrystals, complex geometric patterns: Peter J. Lu and Paul J. Steinhardt, "Decagonal and Quasi-Crystalline Tilings in Medieval Islamic Architecture," *Science* 315, no. 5815 (February 23, 2007): 1106–10, https://doi.org/10.1126/science.1135491.

a similar technique called LiDAR: Leah A. Wasser, "The Basics of LiDAR," National Ecological Observatory Network, updated September 13, 2024, https://www.neonscience.org/resources/learning-hub/tutorials/lidar-basics; "What Is Lidar?," National Oceanic and Atmospheric Administration, accessed February 20, 2025, https://oceanservice.noaa.gov/facts/lidar.html.

In an essay published: The original 1907 essay is not readily available, but it was reprinted in Bertrand Russell, *Philosophical Essays* (London: Longmans, Green, and Co., 1910), 73.

Slight fluctuations in the density: Leembruggen, discussion, May 10, 2024; Leembruggen, email, December 3 and 11, 2024; Rapidis, email, October 31 and November 5, 2024.

artwork by seventeenth-century Dutch: "Painting in the Dutch Golden Age: A Profile of the Seventeenth Century," National Gallery of Art, 2007, https://www.nga.gov/content/dam/ngaweb/Education/learning-resources/teaching-packets/pdfs/dutch_painting.pdf.

Molecular Biology: Bright Lines

In December 1832: John Snow, *On the Mode of Communication of Cholera*, 2nd ed. (London: John Churchill, 1855), 4–5, https://wellcomelibrary.org/item/b28985266.

In another town, residents: Snow, *On the Mode*, 32.

Snow's most famous contribution: William Bynum, "In Retrospect: On the Mode of Communication of Cholera," *Nature* 495 (2013): 169–70, https://doi.org/10.1038/495169a; Snow, *On the Mode*, 38–40.

nine customers of a coffee shop: Snow, *On the Mode*, 41–42.

twenty-eight-year-old pregnant woman: Snow, *On the Mode*, 43.

man who traveled from Brighton: Snow, *On the Mode*, 44.
the percussion–cap maker's wife: Snow, *On the Mode*, 44.
the tailor, the army officer: Snow, *On the Mode*, 43, 47.
It was, Snow wrote: Snow, *On the Mode*, 38.
Snow kept meticulous tables: Snow, *On the Mode*, 49, 62, 71, 73, 84, 85, 138–62. The map is not on a numbered page but appears between pp. 44 and 45; the explanation of the map is on the page after p. vii (not numbered) and pp. 45–46; also see Tom Koch and Ken Denike, "Essential, Illustrative, or… Just Propaganda? Rethinking John Snow's Broad Street Map," *Cartographica: The International Journal for Geographic Information and Geovisualization* 45, no. 1 (March 2010): 19–31, https://doi.org/10.3138/carto.45.1.19.
In part of his report: Snow, *On the Mode*, 60–92; Bynum, "In Retrospect."
"I resolved to spare": Snow, *On the Mode*, 76.
SARS-CoV-2, the virus: Hangping Yao et al., "Molecular Architecture of the SARS-CoV-2 Virus," *Cell* 183, no. 3 (October 2020): 730–38.e13, https://doi.org/10.1016/j.cell.2020.09.018.
scientists around the world: Amy Maxmen, "One Million Coronavirus Sequences: Popular Genome Site Hits Mega Milestone," *Nature*, April 23, 2021, https://www.nature.com/articles/d41586-021-01069-w; Louise Moncla (virologist, University of Pennsylvania), in discussion with the author, April 23, 2020, and May 13, 2020; Björn Meyer (RNA virologist, Otto von Guericke University Magdeburg), in discussion with the author, December 7, 2020; Cedric C. S. Tan et al., "Transmission of SARS-CoV-2 from Humans to Animals and Potential Host Adaptation," *Nature Communications* 13, no. 1 (May 27, 2022): 2988, https://doi.org/10.1038/s41467-022-30698-6.
Because the genome was mutating: Trevor Bedford, "Cryptic Transmission of Novel Coronavirus Revealed by Genomic Epidemiology," Bedford Lab, March 2, 2020, https://bedford.io/blog/ncov-cryptic-transmission; Moncla, discussion, April 23, 2020; Sidney Bell (computational biologist, Chan Zuckerberg Initiative), in discussion with the author, August 12, 2020.
the virus's genetic code: The virus genome is in the form of RNA, which is made of the nucleotides adenine (A), uracil (U), cytosine (C), and guanine (G). However, when scientists record SARS-CoV-2 genome sequences, they typically list all the U's as T's instead. This is because biologists convert the virus's RNA into DNA in order to sequence it; RNA is similar to DNA, but DNA contains thymine (T) instead of uracil (U). Meyer, discussion; Gage Moreno (viral genomicist, Broad Institute), in discussion with the author, March 22, 2022; "Ribonucleic Acid (RNA)," National Human Genome Research Institute, accessed April 22, 2025, https://www.genome.gov/about-genomics/educational-resources/fact-sheets/ribonucleic-acid-fact-sheet; B.

Alberts et al., "From DNA to RNA," in *Molecular Biology of the Cell*, 4th ed. (New York: Garland Science, 2002), https://www.ncbi.nlm.nih.gov/books/NBK26887.

about thirty thousand more letters: Fan Wu et al., "A New Coronavirus Associated with Human Respiratory Disease in China," *Nature* 579, no. 7798 (March 12, 2020): 265–69, https://doi.org/10.1038/s41586-020-2008-3; Sergey Nurk et al., "The Complete Sequence of a Human Genome," *Science* 376 (2022): 44–53, doi:10.1126/science.abj6987; "Human Genomic Variation," National Human Genome Research Institute, accessed April 22, 2025, https://www.genome.gov/about-genomics/educational-resources/fact-sheets/human-genomic-variation; Meyer, discussion.

the death toll from COVID-19: "An Incalculable Loss," *New York Times*, updated May 27, 2020, https://www.nytimes.com/interactive/2020/05/24/us/us-coronavirus-deaths-100000.html; "This Week in Coronavirus: May 21 to May 28," *KFF*, May 29, 2020, https://www.kff.org/policy-watch/this-week-in-coronavirus-may-21-to-may-28.

It had started with: Edward Holmes (@edwardcholmes), "All, an initial genome sequence of the coronavirus associated with the Wuhan outbreak is now available at Virological.org here: https://virological.org/t/novel-2019-coronavirus-genome/319," Twitter, January 11, 2020, 1:08 a.m., archived April 16, 2022, at https://web.archive.org/web/20220416031852/https://twitter.com/edwardcholmes/status/1215802670176276482?s=20; Martin Enserink, "Dispute Simmers over Who First Shared SARS-CoV-2's Genome," *Science*, March 29, 2023, https://www.science.org/content/article/dispute-simmers-over-who-first-shared-sars-cov-2-s-genome; Jemma Geoghegan (evolutionary virologist, University of Otago), in discussion with the author, December 17, 2020.

Edward Holmes, a virologist: The sequence produced by Zhang's team, posted on the website virological.org, was widely hailed as the first publicly available SARS-CoV-2 genome sequence. A data science initiative called GISAID has said that the first SARS-CoV-2 sequences were available in its database earlier than that, but some scientists have disputed this claim. "Novel 2019 Coronavirus Genome," Virological.org, January 10, 2020, https://virological.org/t/novel-2019-coronavirus-genome/319; David Cyranoski, "Zhang Yongzhen: Genome Sharer," *Nature*, December 14, 2020, https://www.nature.com/immersive/d41586-020-03435-6/index.html#zhang-yongzhen; Enserink, "Dispute Simmers"; "GISAID Comments on the Speculations Surrounding Data Availability," GISAID, March 21, 2023, https://gisaid.org/statements-clarifications/data-availability; Talha Burki, "First Shared SARS-CoV-2 Genome: GISAID vs Virological.org," April 25, 2023, doi:10.1016/s2666-5247(23)00133-7.

The sample had been collected: Wu et al., "A New Coronavirus"; Enserink, "Dispute Simmers."

Over the next sixteen months: Maxmen, "One Million Coronavirus Sequences"; Zhiyuan Chen et al., "Global Landscape of SARS-CoV-2 Genomic Surveillance and Data Sharing," *Nature Genetics* 54, no. 4 (April 2022): 499–507, https://doi.org/10.1038/s41588-022-01033-y; Louis du Plessis et al., "Establishment and Lineage Dynamics of the SARS-CoV-2 Epidemic in the UK," *Science* 371, no. 6530 (February 12, 2021): 708–12, https://doi.org/10.1126/science.abf2946; Emma Hodcroft (molecular epidemiologist, University of Basel), in discussion with the author, September 22, 2020.

As those sequences poured: The description of Nextstrain analysis was synthesized from Trevor Bedford, "Early Warnings of Novel Coronavirus from Genomic Epidemiology and the Global Open Scientific Response," Bedford Lab, January 31, 2020, https://bedford.io/blog/genomic-epi-for-ncov-response; Bedford, "Cryptic Transmission of Novel Coronavirus"; "How to Interpret the Phylogenetic Trees," Nextstrain, accessed April 23, 2025, https://docs.nextstrain.org/en/latest/learn/interpret/how-to-read-a-tree.html; "Nextstrain SARS-CoV-2 Resources," Nextstrain, accessed April 23, 2025, https://nextstrain.org/sars-cov-2; "Genomic Epidemiology of SARS-CoV-2 with Subsampling Focused Globally over the Past 6 Months," Nextstrain, accessed April 23, 2025, https://nextstrain.org/ncov/gisaid/global/6m; Moncla, discussion, April 23, 2020; Miguel Paredes (MD-PhD student, University of Washington), in discussion with the author, August 14, 2020; Hodcroft, discussion, November 24, 2020.

In an early report: Bedford et al., "Genomic Analysis of nCoV Spread. Situation Report 2020-01-25," Nextstrain, January 25, 2020, https://nextstrain.org/narratives/ncov/sit-rep/2020-01-25; "Phylogenetic Analysis," Nextstrain, January 25, 2020, https://nextstrain.org/narratives/ncov/sit-rep/2020-01-25?n=5; Trevor Bedford et al., "Cryptic Transmission of SARS-CoV-2 in Washington State," *Science* 370, no. 6516 (October 30, 2020): 571–75, https://doi.org/10.1126/science.abc0523.

Four days later, news outlets: Sydney Brownstone et al., "King County Patient Is First in U.S. to Die of Covid-19 as Officials Scramble to Stem Spread of Novel Coronavirus," *Seattle Times*, February 29, 2020, https://www.seattletimes.com/seattle-news/health/one-king-county-patient-has-died-due-to-covid-19-infection; Mike Baker, Nicholas Bogel-Burroughs, and Karen Weise, "Washington State Declares Emergency Amid Coronavirus Death and Illnesses at Nursing Home," *New York Times*, February 29, 2020, https://www.nytimes.com/2020/02/29/us/coronavirus-washington-death.html; Eric Boodman and Helen Branswell, "First Covid-19 Outbreak in a U.S. Nursing Home Raises

Concerns," *STAT*, February 29, 2020, https://www.statnews.com/2020/02/29/new-covid-19-death-raises-concerns-about-virus-spread-in-nursing-homes.

When it became clear: Mike Baker et al., "2nd Death Near Seattle Adds to Signs Virus Is Spreading in U.S.," *New York Times*, March 1, 2020, https://www.nytimes.com/2020/03/01/us/washington-coronavirus-nursing-home.html.

I heard about the number: Moncla, discussion, April 23, 2020, and May 13, 2020; Cassia Wagner (MD-PhD student, University of Washington), in discussion with the author, July 30, 2020; Moira Zuber (master's student, University of Basel), in discussion with the author, October 7, 2020; Hodcroft, in discussion with the author, November 24, 2020; Maxmen, "One Million Coronavirus Sequences."

They told me about cases: Moreno, discussion; Paraic Kenny (director, Gundersen Medical Foundation's Kabara Cancer Research Institute), in discussion with the author, November 23, 2020; Geoghegan, in discussion with the author, June 3, 2021.

"In December 2019, a new": Trevor Bedford et al., "Recent Outbreak of a Novel Coronavirus," Nextstrain, January 23, 2020, https://nextstrain.org/narratives/ncov/sit-rep/2020-01-23?n=3.

Miguel Paredes's medical school class: Paredes, discussion, August 14, 2020.

Bedford's team and their collaborators: James Hadfield et al., "Nextstrain: Real-Time Tracking of Pathogen Evolution," ed. Janet Kelso, *Bioinformatics* 34, no. 23 (December 1, 2018): 4121–23, https://doi.org/10.1093/bioinformatics/bty407; Moncla, discussion, April 23, 2020; Hodcroft, discussion, September 22, 2020.

After moving to his parents': Paredes, discussion, August 14, 2020.

The way that Nextstrain: I've used DNA nucleotides (A, T, C, G) in this explanation rather than RNA nucleotides (A, U, C, G), because these are the letters that biologists typically use even when discussing RNA sequences. Nextstrain is one of many software options for performing this type of analysis on genome sequences, but the main difference is that Nextstrain brings together a set of analysis tools into one "pipeline" and generates interactive visualizations of the results. The paragraphs about the telephone analogy were synthesized from Bell, discussion; Hadfield et al., "Nextstrain: Real-Time Tracking"; Bedford et al., "Cryptic Transmission of SARS-CoV-2"; Bedford, "Cryptic Transmission of Novel Coronavirus"; Moncla, discussion, April 23, 2020; Paredes, discussion, August 14, 2020.

The picture was still incomplete: Geoghegan, discussion, December 17, 2020; Wagner, discussion; Moncla, discussion, April 23, 2020, and November 25, 2020; Bell, discussion.

The practice of reconstructing disease: Graham Mooney (historian, Johns Hopkins School of Medicine), in discussion with the author, February 3, 2021; Samuel Cohn (historian, University of Glasgow), in discussion with the author, January 25, 2021; Graham Mooney, *Intrusive Interventions: Public Health, Domestic Space, and Infectious Disease Surveillance in England, 1840–1914* (University of Rochester Press, 2015), 110.

In a public health report: The description of Nellie and the measles outbreak appears in an appendix written by Dr. Thomas, which is part of James Kerr, "Report of the Medical Officer (Education) for the Year Ended 31st March, 1905," in *Report of the Public Health Committee of the London County Council, Submitting the Report of the Medical Officer of Health of the County for the Year 1904* (London County Council, 1905): 46–60, https://wellcomecollection.org/works/tdrkjgu9. Nellie's case is described in detail on pp. 51–52; the quote appears on p. 52. The case is also described in Mooney, *Intrusive Interventions*, 110–11.

The first trees were neat: Trevor Bedford et al., "Genomic Analysis of nCoV Spread. Situation Report 2020-01-23: Phylogenetic Analysis," Nextstrain, January 23, 2020, https://nextstrain.org/narratives/ncov/sit-rep/2020-01-23?n=5; Trevor Bedford et al., "Genomic Analysis of nCoV Spread. Situation Report 2020-01-25: Phylogenetic Analysis," Nextstrain, January 25, 2020, https://nextstrain.org/narratives/ncov/sit-rep/2020-01-25?n=5.

Then one version of SARS-CoV-2: Trevor Bedford et al., "Genomic Analysis of COVID-19 Spread. Situation Report 2020-03-04," Nextstrain, March 4, 2020, https://nextstrain.org/narratives/ncov/sit-rep/2020-03-04. Click through to the pages "'Divergence' in Phylogenies" and "Possible Hidden Transmission in Italy."

The number of sequences climbed: The growing number of sequences in Nextstrain trees can be seen by viewing the chronologically listed situation reports at "Nextstrain SARS-CoV-2 Resources," under the heading "All SARS-CoV-2 situation reports."

By late March: Sidney M. Bell et al., "Genomic Analysis of COVID-19 Spread. Situation Report 2020-03-27," Nextstrain, March 27, 2020, https://nextstrain.org/narratives/ncov/sit-rep/2020-03-27.

During the early days: The description of Moreno and Braun's work was synthesized from Moreno, discussion; Katarina Braun (resident in obstetrics, gynecology, and reproductive sciences, Yale New Haven Health), in discussion with the author, March 31, 2022; Gage K. Moreno et al., "Revealing Fine-Scale Spatiotemporal Differences in SARS-CoV-2 Introduction and Spread," *Nature Communications* 11, no. 1 (December 2020): 5558, https://doi.org/10.1038/s41467-020-19346-z.

Once they'd prepared the DNA: Moreno, discussion; "How Nanopore

Sequencing Works," Oxford Nanopore Technologies, May 29, 2019, YouTube video, 1:41, https://www.youtube.com/watch?v=RcP85JHLmnI.

A couple hours to the west: Kenny, discussion, November 23, 2020, and May 27, 2021; Craig Richmond (lead research technician, Gundersen Medical Foundation's Kabara Cancer Research Institute), in discussion with the author, December 9, 2020.

The researchers wondered: The college student and nursing home study is described in Craig S. Richmond et al., "SARS-CoV-2 Sequencing Reveals Rapid Transmission from College Student Clusters Resulting in Morbidity and Deaths in Vulnerable Populations," medRxiv preprint (October 14, 2020), https://doi.org/10.1101/2020.10.12.20210294.

Nouar's team wanted to investigate: Nouar Qutob (geneticist, Arab American University), in discussion with the author, January 13, 2021.

the team confirmed: Nouar Qutob et al., "Genomic Epidemiology of the First Epidemic Wave of Severe Acute Respiratory Syndrome Coronavirus 2 (SARS-CoV-2) in Palestine," *Microbial Genomics* 7, no. 6 (June 22, 2021), https://doi.org/10.1099/mgen.0.000584. The cases at the hotel are described in Adam Rasgon, "PA: 4 People at Bethlehem-Area Hotel May Have Virus, Nativity Church to Shutter," *Times of Israel*, March 5, 2020, https://www.timesofisrael.com/pa-4-people-at-bethlehem-area-hotel-may-have-virus-nativity-church-to-shutter; Majeda El Batsh, "Palestinians Confirm 7 Coronavirus Cases, Declare Tourist Ban," *Times of Israel*, March 5, 2020, https://www.timesofisrael.com/palestinians-confirm-7-coronavirus-cases-declare-tourist-ban; Ewan Palmer, "15 Americans Held in West Bank Hotel Amid Fears of Coronavirus Infection," *Newsweek*, March 7, 2020, https://www.newsweek.com/coronavirus-tourists-hotel-palestine-bethlehem-1491046.

Victoria, the scientist in Uruguay: The description of the work in Uruguay and New York was synthesized from Maria Victoria Elizondo (biotechnologist, Asociación Española Primera en Salud), in discussion with the author, December 11, 2020, and July 21, 2021; Dacia Dimartino (molecular biologist, New York University Langone Health's Genome Technology Center), in discussion with the author, May 25, 2022; Adriana Heguy (director, New York University Langone Health's Genome Technology Center), in discussion with the author, December 1, 2021.

After sequencing forty-four samples: Victoria Elizondo et al., "SARS-CoV-2 Genomic Characterization and Clinical Manifestation of the COVID-19 Outbreak in Uruguay," *Emerging Microbes & Infections* 10, no. 1 (January 1, 2021): 51–65, https://doi.org/10.1080/22221751.2020.1863747.

news reports were blowing up: Uki Goñi, "Half of Uruguay's Coronavirus Cases Traced to a Single Guest at a Society Party," *Guardian*, March 19, 2020, https://www.theguardian.com/world/2020/mar/19/uruguay-coronavirus-party-guest-argentina; Sebastián Fest, "'¡Están Quemando Mi Marca, Creen Que Soy Una Terrorista!': El Descargo De Carmela Hontou, La Diseñadora Uruguaya Que Contrajo Coronavirus Y Fue a Un Casamiento," *Infobae*, March 15, 2020, https://www.infobae.com/america/america-latina/2020/03/15/estan-quemando-mi-marca-creen-que-soy-una-terrorista-el-descargo-de-carmela-hontou-la-disenadora-uruguaya-que-contrajo-coronavirus-y-violo-la-cuarentena; Dan Adler, "A Society Wedding Is at the Center of Uruguay's Covid-19 Outbreak," *Vanity Fair*, March 20, 2020, https://www.vanityfair.com/style/2020/03/society-wedding-uruguay-covid-19-outbreak.

a case in sixteenth-century Italy: The description of the plague cases in Italy was synthesized from Andrea Gratiolo di Salò, *Discorso di Peste* (1576), 28–29, https://www.google.com/books/edition/Discorso_di_peste_di_M_Andrea_Gratiolo_d/8YzyUuS9W0YC?hl=en; Samuel K. Cohn Jr., "Plague Disputes, Challenges of the 'Universals'" in *Cultures of Plague: Medical Thinking at the End of the Renaissance* (Oxford University Press, 2009), 32, https://academic.oup.com/book/34307/chapter-abstract/291011920?redirectedFrom=fulltext; Samuel Cohn and Mona O'Brien, "Contact Tracing: How Physicians Used It 500 Years Ago to Control the Bubonic Plague," *The Conversation*, June 3, 2020, https://theconversation.com/contact-tracing-how-physicians-used-it-500-years-ago-to-control-the-bubonic-plague-139248; Cohn, discussion; Cohn, email messages to fact-checker, April 23, 2025, and May 19, 2025.

Placide Mbala-Kingebeni was accustomed: Information about the DRC team's sequencing work during Ebola outbreaks was synthesized from Placide Mbala-Kingebeni (head of epidemiology and global health department, Institut National de Recherche Biomédicale), in discussion with the author, January 11, 2021; Eddy Kinganda-Lusamaki (virologist, Institut National de Recherche Biomédicale), in discussion with the author, May 25, 2022; Eddy Kinganda-Lusamaki et al., "Integration of Genomic Sequencing into the Response to the Ebola Virus Outbreak in Nord Kivu, Democratic Republic of the Congo," *Nature Medicine* 27, no. 4 (April 2021): 710–16, https://doi.org/10.1038/s41591-021-01302-z; Eddy Kinganda-Lusamaki et al., "2020 Ebola Virus Disease Outbreak in Équateur Province, Democratic Republic of the Congo: A Retrospective Genomic Characterisation," *The Lancet Microbe* 5, no. 2 (February 2024): e109–18, https://doi.org/10.1016/S2666–5247(23)00259–8.

They had to contend: Details about local unrest and the MSF team were synthesized from Chad R. Wells et al., "The Exacerbation of Ebola Outbreaks by Conflict in the Democratic Republic of the Congo," *Proceedings of the*

National Academy of Sciences 116, no. 48 (November 26, 2019): 24366–72, https://doi.org/10.1073/pnas.1913980116 (particularly the supplementary information, which includes a narrative account by Bernard Gaüzère, senior Ebola MD with MSF in Katwa); "DRC: MSF Shuts Down Ebola Treatment Center Following Violent Attack," Médecins Sans Frontières, February 26, 2019, https://www.doctorswithoutborders.org/latest/drc-msf-shuts-down-ebola-treatment-center-following-violent-attack; "2019 Country Reports on Human Rights Practices: Democratic Republic of the Congo," U.S. Department of State, accessed April 23, 2025, https://www.state.gov/reports/2019-country-reports-on-human-rights-practices/democratic-republic-of-the-congo; Rebecca Ratcliffe, "Arsonists Attack Ebola Clinics in DRC as Climate of Distrust Grows," *Guardian*, February 28, 2019, https://www.theguardian.com/global-development/2019/feb/28/arsonists-attack-ebola-clinics-in-drc-as-climate-of-distrust-grows; Kinganda-Lusamaki et al., "Integration of Genomic Sequencing"; Kinganda-Lusamaki, discussion.

These restrictions, the team wrote: Kinganda-Lusamaki et al., "Integration of Genomic Sequencing," 715.

Nevertheless, they'd gotten results: Kinganda-Lusamaki et al., "Integration of Genomic Sequencing"; Kinganda-Lusamaki, discussion; Allison Black (postdoctoral researcher, Fred Hutchinson Cancer Center), in discussion with the author, December 3, 2020.

His team had started sequencing: Details about the team's SARS-CoV-2 sequencing were synthesized from Mbala-Kingebeni, discussion; Kinganda-Lusamaki, discussion; Amuri Aziza (biologist, Institut National de Recherche Biomédicale), in discussion with the author, January 20, 2021.

At first, the culprit: The description of the quarantine hotel case study was synthesized from Nick Eichler et al., "Transmission of Severe Acute Respiratory Syndrome Coronavirus 2 during Border Quarantine and Air Travel, New Zealand (Aotearoa)," *Emerging Infectious Diseases* 27, no. 5 (May 2021): 1274–78, https://doi.org/10.3201/eid2705.210514; Jordan Douglas et al., "Real-Time Genomics for Tracking Severe Acute Respiratory Syndrome Coronavirus 2 Border Incursions after Virus Elimination, New Zealand," *Emerging Infectious Diseases* 27, no. 9 (September 2021): 2361–68, https://doi.org/10.3201/eid2709.211097; Geoghegan, discussion, December 17, 2020, and June 3, 2021; Sarah Berger (nursing director, Canterbury District Health Board's infection prevention and control service), in discussion with the author, March 25, 2022; Joshua Freeman (acting clinical director of infection prevention and control, Canterbury District Health Board), in discussion with the author, March 22, 2022.

Jemma's team had been: Jemma L. Geoghegan et al., "Genomic Epidemiology

Reveals Transmission Patterns and Dynamics of SARS-CoV-2 in Aotearoa New Zealand," *Nature Communications* 11, no. 1 (December 2020): 6351, https://doi.org/10.1038/s41467-020-20235-8; Geoghegan, discussion, December 17, 2020; "Managing COVID-19 in New Zealand," New Zealand Ministry of Health, accessed April 23, 2025, https://www.health.govt.nz/strategies-initiatives/programmes-and-initiatives/covid-19/managing-covid-19-in-new-zealand.

As one scientist put it: Freeman, discussion.

The case of the quarantine hotel: The paragraphs about the quarantine hotel residents and initial investigation were synthesized from Eichler et al., "Transmission of Severe Acute"; Douglas et al., "Real-Time Genomics for Tracking"; Geoghegan, discussion, December 17, 2020, and June 3, 2021; Berger, discussion; Freeman, discussion.

At the time, the standard: "Transmission of SARS-CoV-2: Implications for Infection Prevention Precautions," World Health Organization (WHO), July 9, 2020, https://www.who.int/news-room/commentaries/detail/transmission-of-sars-cov-2-implications-for-infection-prevention-precautions. Descriptions of the WHO's early stance on transmission can also be found in Dyani Lewis, "Why the WHO Took Two Years to Say COVID Is Airborne," *Nature*, April 6, 2022, https://www.nature.com/articles/d41586-022-00925-7; Jose L. Jimenez et al., "What Were the Historical Reasons for the Resistance to Recognizing Airborne Transmission during the COVID-19 Pandemic?," *Indoor Air* 32, no. 8 (August 2022), https://doi.org/10.1111/ina.13070; Apoorva Mandavilli, "The Coronavirus Can Be Airborne Indoors, W.H.O. Says," *New York Times*, July 9, 2020, https://www.nytimes.com/2020/07/09/health/virus-aerosols-who.html.

Despite urging from many scientists: The WHO said that airborne transmission could happen during aerosol-generating procedures in medical settings. Outside of healthcare facilities, aerosol transmission in some crowded indoor places with poor ventilation "cannot be ruled out" but "requires further study," according to a WHO scientific brief in July 2020. In October 2020, the agency updated its website to say that aerosol transmission "can occur" in certain indoor settings where an infected person has been present for a long time, but still maintained that droplets between people in close contact were the primary mode of transmission. "Transmission of SARS-CoV-2"; "Coronavirus Disease (COVID-19): How Is It Transmitted?," WHO, last modified October 20, 2020, archived April 26, 2021, at https://web.archive.org/web/20210426045337/https:/www.who.int/news-room/q-a-detail/coronavirus-disease-covid-19-how-is-it-transmitted; Lidia Morawska and Donald K Milton, "It Is Time to Address Airborne Transmission of Coronavirus Disease 2019 (COVID-19)," *Clinical Infectious Diseases*, July 6, 2020, ciaa939, https://doi.org/10.1093/cid/ciaa939; Apoorva Mandavilli, "239 Experts with

One Big Claim: the Coronavirus Is Airborne," *New York Times*, July 4, 2020, https://www.nytimes.com/2020/07/04/health/239-experts-with-one-big-claim-the-coronavirus-is-airborne.html; Lewis, "Why the WHO"; Jimenez et al., "Historical Reasons for the Resistance"; Jason Chen et al., "Airborne Transmission: A New Paradigm with Major Implications for Infection Control and Public Health," *New Zealand Medical Journal* 136, no. 1570 (February 17, 2023): 69–77, https://doi.org/10.26635/6965.6028; Julian W. Tang et al., "Covid-19 Has Redefined Airborne Transmission," *BMJ*, April 14, 2021, n913, https://doi.org/10.1136/bmj.n913.

But Berger's team wasn't happy: The description of the final stages of the investigation was synthesized from Berger, discussion; Freeman, discussion; Geoghegan, discussion, June 3, 2021; Eichler et al., "Transmission of Severe Acute"; Douglas et al., "Real-Time Genomics for Tracking"; Chen et al., "Airborne Transmission: A New Paradigm."

the resistance that John Snow met: Details about prevailing views about cholera during Snow's time were synthesized from Stephen Halliday, "Death and Miasma in Victorian London: An Obstinate Belief," *BMJ* 323, no. 7327 (December 22, 2001): 1469–71, https://doi.org/10.1136/bmj.323.7327.1469; Kathleen A. Fairman, "Pandemics, Policy, and the Power of Paradigm: Will COVID-19 Lead to a New Scientific Revolution?," *Annals of Epidemiology* 69 (May 2022): 17–23, https://doi.org/10.1016/j.annepidem.2022.02.005; Nigel Paneth et al., "A Rivalry of Foulness: Official and Unofficial Investigations of the London Cholera Epidemic of 1854," *American Journal of Public Health* 88, no. 10 (October 1998): 1545–53; Steven Johnson, *The Ghost Map: The Story of London's Most Terrifying Epidemic–and How It Changed Science, Cities, and the Modern World* (New York: Riverhead Books, 2006), 68–69, 121–23.

One physician wrote: E. A. Parkes, "*Mode of Communication of Cholera*. By John Snow, M.D. Second Edition," *British and Foreign Medico-Chirurgical Review* 15 (1855): 454, 456, 459, https://pmc.ncbi.nlm.nih.gov/articles/PMC5184364; also see B. P. Bergman, "Commentary: Edmund Alexander Parkes, John Snow and the Miasma Controversy," *International Journal of Epidemiology* 42, no. 6 (December 1, 2013): 1562–65, https://doi.org/10.1093/ije/dyt212.

An editorial in *The Lancet*: Editorial, *The Lancet* 65, no. 1660 (June 23, 1855): 635, https://www.sciencedirect.com/science/article/abs/pii/S0140673602444686; also see Sandra Hempel, "John Snow," *The Lancet* 381, no. 9874 (April 2013): 1269–70, https://doi.org/10.1016/S0140–6736(13)60830–2.

About a decade later: Halliday, "Death and Miasma"; Johnson, *The Ghost Map*, 209–13.

In the late 1800s: Jimenez et al., "Historical Reasons for the Resistance"; Johnson, *The Ghost Map*, 213.

But scientists eventually became reluctant: Jimenez et al., "Historical Reasons for the Resistance," 2.

A key figure was Charles Chapin: Jimenez et al., "Historical Reasons for the Resistance"; George Rosen, "Charles V. Chapin and the Public Health Movement," *American Journal of Public Health and the Nation's Health*, 53, no. 5 (May 1963): 844–45, https://pmc.ncbi.nlm.nih.gov/articles/PMC1254127; Charles V. Chapin, *The Sources and Modes of Infection* (New York: John Wiley & Sons, 1910), 264.

The number of SARS-CoV-2 sequences: David VanInsberghe et al., "Recombinant SARS-CoV-2 Genomes Circulated at Low Levels over the First Year of the Pandemic," *Virus Evolution* 7, no. 2 (December 16, 2021): veab059, https://doi.org/10.1093/ve/veab059.

The software displayed only: Zuber, discussion; Hodcroft, discussion, November 24, 2020.

When Snow had presented: Snow, *On the Mode*, 40; Sidney Chave, "Henry Whitehead and Cholera in Broad Street," *Medical History* 2, no. 2 (April 1958): 92–108, https://doi.org/10.1017/S0025727300023504.

Within a week or so: Snow, *On the Mode*, 49, 51; Henry Whitehead, "Remarks on the Outbreak of Cholera in Broad Street, Golden Square, London, 1854," *Transactions of the Epidemiological Society of London* 3, no. 1 (1869): 99–104, https://pmc.ncbi.nlm.nih.gov/articles/PMC5523601.

The New Zealand team's investigation: Berger, discussion; Freeman, discussion; Chen et al., "Airborne Transmission: A New Paradigm."

Katarina and Gage found: Moreno, discussion.

By the time his results: Richmond et al., "SARS-CoV-2 Sequencing Reveals"; Kenny, email message to fact-checker, January 15, 2025; Maija Sikora, "UWL Coate Hall Under 'Shelter-In-Place' Order; Students Told Not to Leave Their Rooms," *Racquet Press*, September 11, 2020, https://theracquet.org/10486/news/uwl-coate-hall-under-shelter-in-place-order-students-told-not-to-leave-their-rooms; Julia Balli, "UWL Announces All Classes Online Until Sept. 28, Immediate Shelter in Place for All Residence Halls," *Racquet Press*, September 13, 2020, https://theracquet.org/10506/news/uwl-announces-all-classes-online-until-sept-28-immediate-shelter-in-place-for-all-residence-halls.

Dave Eggers's novel *The Every*: Dave Eggers, *The Every* (Vintage, 2021), 6.

Removing the pump handle: Snow, *On the Mode*, 38, 51; Nigel Paneth, "Assessing the Contributions of John Snow to Epidemiology: 150 Years After Removal of the Broad Street Pump Handle," *Epidemiology* 15, no. 5 (September 2004): 514–16, https://doi.org/10.1097/01.ede.0000135915

.94799.00; Whitehead, "Remarks on the Outbreak"; Chave, "Henry Whitehead and Cholera"; Cholera Inquiry Committee, *Report on the Cholera Outbreak in the Parish of St. James, Westminster, During the Autumn of 1854* (London: J. Churchill, 1855), 153, https://wellcomecollection.org/works/z8xczc2r.

In his book *The Ghost Map*: Johnson, *The Ghost Map*, 162–63.

The ripple effects: Details about the case of the sick baby were synthesized from Whitehead, "Remarks on the Outbreak"; Chave, "Henry Whitehead and Cholera"; Cholera Inquiry Committee, *Report on the Cholera*, 159–163; Henry Whitehead, "The Broad Street Pump: An Episode in the Cholera Epidemic of 1854," *Macmillan's Magazine* (December 1865): 113–22, https://www.google.com/books/edition/Macmillan_s_Magazine/RvpAAQAAMAAJ?hl=en; Johnson, *The Ghost Map*, 187–88.

This chain of events: Whitehead, "The Broad Street Pump"; Chave, "Henry Whitehead and Cholera"; Cholera Inquiry Committee, *Report on the Cholera*, 138–39; Johnson, *The Ghost Map*, 167–68.

Whitehead was confident: Whitehead, "The Broad Street Pump"; Cholera Inquiry Committee, *Report on the Cholera*, 133, 139, 162. The quote is from Cholera Inquiry Committee, *Report on the Cholera*, 125.

His detailed investigation helped convince: Chave, "Henry Whitehead and Cholera"; Cholera Inquiry Committee, *Report on the Cholera*, 76, 81; Johnson, *The Ghost Map*, 199–201. The quote is from Cholera Inquiry Committee, *Report on the Cholera*, 83; also see pp. 87–91 for other hypotheses the committee considered.

He told Whitehead: "The Experience of a London Curate: Reprint of Farewell Speech at the Rainbow Tavern," in H. D. Rawnsley, *Henry Whitehead (1825–1896): A Memorial Sketch* (Glasgow: James MacLehose and Sons, 1898), 205–6, https://babel.hathitrust.org/cgi/pt?id=nyp.33433104209287&seq=224; Johnson, *The Ghost Map*, 181.

more than two million people: "Weekly Epidemiological Update—23 February 2021," WHO, February 23, 2021, https://www.who.int/publications/m/item/weekly-epidemiological-update---23-february-2021.

it was, as Snow said: "Experience of a London Curate," 206.

Chemistry: A Space of Possibility

Details about Richelle Thomas's research were drawn from interviews with Thomas and her colleagues Robert Root, Gilberto Curlango-Rivera, and Jean McLain, and from my reporting trip to the University of Arizona and the Navajo Nation. Details about research in Jani Ingram's lab were drawn from interviews with Ingram; former and current lab members Jonathan Credo, Andee Lister, Marissa Mares, Tommy Rock, Colleen Cooley, Lydia Edgewater, Emily Werner, and Shanadeen Begay; and collaborators Joe Hoover and Kevin Webster. General background about uranium mining and its effects on the community were drawn from interviews with Perry Charley, Candis Yazzie, Esther Yazzie-Lewis, Douglas Brugge, Rob Goble, Chris Shuey, Johnnye Lewis, Margaret Briehl, Anita Moore-Nall, Michelle David, and Eric Jantz. Interviews with community members Lula Neztsosie, Ray Yellowhorse, and Nolan Stevens were conducted during my reporting trip to Cameron.

a small community: "Cameron CDP, Arizona," United States Census Bureau, accessed April 16, 2025, https://data.census.gov/profile?g=160XX00US0409340.

more than 25,000 square miles: "Navajo Area," Indian Health Service, https://www.ihs.gov/navajo, accessed April 16, 2025; "Navajo Nation Department of Agriculture," Yá'át'ééh: The Official Site of the Navajo Nation, https://agriculture.navajo-nsn.gov, accessed April 16, 2025.

where at least 150,000: My visit was in 2023. Obtaining an accurate population count is difficult because the US Census Bureau has undercounted Indigenous people living on tribal land. According to 2020 Census data, about 165,000 people total lived on the Navajo Nation, and roughly 158,000 of them identified as only "American Indian and Alaska Native": "Navajo Nation Reservation and Off-Reservation Trust Land, AZ—NM—UT," United States Census, https://data.census.gov/profile/Navajo_Nation_Reservation_and_Off-Reservation_Trust_Land,_AZ—NM—UT?g=2500000US2430, accessed April 16, 2025. According to a 2024 report by the Navajo Epidemiology Center, based on data from the 2021 American Community Survey, about 159,000 people categorized as "Navajo alone or in combination" lived on the Navajo Nation: "Navajo Nation Population Profile: U.S. Census 2020," Navajo Epidemiology Center, July 2024, https://nec.navajo-nsn.gov/Portals/0/Reports/Navajo%20Nation%20Population%20Profile%202020.pdf. The Navajo Nation's Department of Agriculture website states that 168,000 enrolled tribal members live on the Navajo Nation: "Navajo Nation Department of Agriculture," Yá'át'ééh: The Official Site of the Navajo Nation, accessed July 24, 2025, https://agriculture.navajo-nsn.gov. In a 2022 media interview, Jonathan Nez, then president of the Navajo Nation, was quoted saying that 180,000 enrolled members lived on the Navajo Nation: Shondiin Silversmith, "Large Census Undercount of Indigenous People on Tribal Lands Means Fewer Resources, Political Power," *Arizona Mirror*, April 7, 2022, https://azmirror.com/2022/04/07/large-census-undercount-of-indigenous-people-on-tribal-lands-means-fewer-resources-political-power.

intensive uranium mining that had: Doug Brugge and Rob Goble, "The History of Uranium Mining and the Navajo People," *American Journal of Public Health* 92, no. 9 (September 2002): 1410–19, https://doi.org/10.2105/AJPH.92.9.1410; Peter H. Eichstaedt, *If You Poison Us: Uranium and Native Americans* (Red Crane Books, 1994), 33–36; "Abandoned Mines Cleanup," U.S. Environmental Protection Agency, accessed April 16, 2025, https://www.epa.gov/navajo-nation-uranium-cleanup/aum-cleanup.

arsenic—another element present: Jonathan Credo et al., "Quantification of Elemental Contaminants in Unregulated Water across Western Navajo Nation," *International Journal of Environmental Research and Public Health* 16, no. 15 (July 31, 2019): 2727, https://doi.org/10.3390/ijerph16152727.

Jani grew up: Details about Jani's life were synthesized from Jani Ingram (chemist, Northern Arizona University), in discussion with the author, May 1, 2017, and September 19, 2022; Jani C. Ingram, "How Motherhood Shaped My Professorship," in *Mom the Chemistry Professor: Personal Accounts and Advice from Chemistry Professors Who Are Mothers*, 2nd ed., eds. Kimberly Woznack et al. (Springer, 2018), 209–12.

Uranium mining had taken off: Information about the ramp-up of uranium mining on the Navajo Nation was synthesized from Brugge and Goble, "History of Uranium Mining"; Eichstaedt, *If You Poison Us*, 9, 31–36, 67; Michelle David, "Clean Up Your Act: The U.S. Government's CERCLA Liability for Uranium Mines on the Navajo Nation," *The University of Chicago Law Review* 90, no. 6 (2023): 1771–1818, https://chicagounbound.uchicago.edu/uclrev/vol90/iss6/5; Doug Brugge (environmental health researcher, University of Connecticut), in discussion with the author, November 9, 2023; Chris Shuey (environmental health scientist, Southwest Research and Information Center), in discussion with the author, November 15, 2023.

For Diné men: Doug Brugge, Timothy Benally, and Esther Yazzie-Lewis, eds., *The Navajo People and Uranium Mining* (University of New Mexico Press, 2006), xvi, xvii, 15; Brugge and Goble, "History of Uranium Mining"; Brugge, discussion; Esther Yazzie-Lewis (Navajo linguist and board member, Southwest Research and Information Center), in discussion with the author, November 8, 2023.

the government was already aware: Brugge and Goble, "History of Uranium Mining"; Duncan A. Holaday, Wilfred D. David, and Henry N. Doyle, "An Interim Report of a Health Study of the Uranium Mines and Mills," May 1952, National Library of Medicine, https://findingaids.nlm.nih.gov/repositories/4/digital_objects/35977; Joseph K. Wagoner et al., "Cancer Mortality Patterns Among U.S. Uranium Miners and Millers, 1950 through 1962," *Journal of the National Cancer Institute* 32, no. 4 (April 1964): 787–801, https://doi.org/10.1093/jnci/32.4.787; Eichstaedt, *If You Poison Us*, 47, 54.

But many Navajo miners: Brugge and Goble, "History of Uranium Mining"; Brugge, Benally, and Yazzie-Lewis, *Navajo People*, 14, 80–81, 131; Eichstaedt, *If You Poison Us*, 97–98.

In the 1950s: Brugge and Goble, "History of Uranium Mining"; Eichstaedt, *If You Poison Us*, 60–64.

feared that knowledge: Perry H. Charley (professor emeritus, Diné College), in discussion with the author, November 7, 2023; Eichstaedt, *If You Poison Us*, 62–63; Brugge and Goble, "History of Uranium Mining."

By the early 1950s: Details about uranium decay and cancer were synthesized from Rob Goble (environmental scientist trained in particle physics, Clark University), in discussion with the author, November 6, 2023; Robert Root (environmental geochemist, University of Arizona), in discussion with the author, March 7, 2023; Brugge, discussion; Margaret Briehl (cancer biologist, University of Arizona), in discussion with the author, November 14, 2023; "Radioactive Decay," U.S. Environmental Protection Agency, accessed April 16, 2025, https://www.epa.gov/radiation/radioactive-decay; Ingmar Grenthe et al., "Uranium," in *Chemistry of the Actinide and Transactinide Elements*, eds. Lester R. Morss, Norman M. Edelstein, and Jean Fuger (Springer, 2011), 256; Brugge and Goble, "History of Uranium Mining"; Eichstaedt, *If You Poison Us*, 49; Holaday, "An Interim Report."

Ventilation in the mines: Eichstaedt, *If You Poison Us*, 79, 81, 92; Brugge and Goble, "History of Uranium Mining."

It wasn't just the miners: Details about the effects on families were synthesized from Yazzie-Lewis, discussion; Brugge, Benally, and Yazzie-Lewis, *Navajo People*, 51–53.

In the 1960s and 1970s: Brugge and Goble, "History of Uranium Mining"; Brugge, Benally, and Yazzie-Lewis, *Navajo People*, xvii, 58; Eichstaedt, *If You Poison Us*, 95–97; Briehl, discussion; Yazzie-Lewis, discussion.

Some Indian Health Service: Johnnye Lewis (inhalation toxicologist, University of New Mexico), in discussion with the author, November 27, 2023.

Few Diné men: Brugge and Goble, "History of Uranium Mining"; M. K. Schubauer-Berigan, R. D. Daniels, and L. E. Pinkerton, "Radon Exposure and Mortality Among White and American Indian Uranium Miners: An Update of the Colorado Plateau Cohort," *American Journal of Epidemiology* 169, no. 6 (January 6, 2009): 718–30, https://doi.org/10.1093/aje/kwn406.

Tommy Rock, a former: The story of Rock's grandfather was synthesized from Tommy Rock (environmental scientist, Northern Arizona University), in discussion with the author, March 24, 2023; Tommy Rock, "Uranium Mining and My Family's Story," *Outrider*, July 22, 2019, https://outrider.org/nuclear-weapons/articles/uranium-mining-and-my-familys-story.

In 1966, a federal government: The development freeze was put in place purportedly to stop the Navajo and Hopi tribes from "taking advantage of the other while they negotiated ownership" of disputed land, according to the Bureau of Indian Affairs. People could not build new houses, schools, businesses, roads, or infrastructure for utilities, apart from water wells and two "administrative safe zones"; residents weren't even allowed to perform home repairs such as fixing roofs. The tribes reached a resolution in 2006, and the freeze was lifted afterward. Information about the freeze was synthesized from "History," Indian Affairs, accessed April 16, 2025, https://www.bia.gov/regional-offices/navajo/western-navajo-agency/environmental-assessment/history; "Final Programmatic Environmental Assessment: Former Bennett Freeze Area Integrated Resource Management Plan," Indian Affairs, September 2021, https://www.bia.gov/sites/default/files/dup/assets/bia/navreg/irmp/2021.09.20_Final_FBFA_IRMP_PEA_508.pdf; "About the Bennett Freeze," Navajo Thaw Implementation Plan, accessed April 16, 2025, https://navajothaw.com/about-the-bennett-freeze; Rima Krisst, "'Enough Talk': Former Bennett Freeze Residents Hope New Initiative Works," *Navajo Times*, February 28, 2020, https://navajotimes.com/reznews/enough-talk-former-bennett-freeze-residents-hope-new-initiative-works; Kate Linthicum, "Trying to Rebuild After 40 Frozen Years," *Los Angeles Times*, November 5, 2009, https://www.latimes.com/la-na-bennet-freeze-5–2009nov5-story.html; Arlyssa D. Becenti, "For Years, They Were Trapped by a Land Dispute. Some Navajo Families Still Wait for Help," *azcentral*, April 15, 2024, https://www.azcentral.com/story/news/local/arizona/2024/04/15/navajo-families-still-waiting-for-help-after-old-federal-land-dispute/73266890007; Candis Yazzie (vice president, Cameron Chapter), in discussion with the author, April 14, 2023.

The Navajo Nation is a sovereign nation: Information about the Navajo Nation and chapter houses was synthesized from Elizabeth Yeager Washington and Stephanie Van Hover, "*Diné Bikéya:* Teaching about Navajo Citizenship and Sovereignty," *The Social Studies* 102, no. 2 (February 7, 2011): 80–87, https://doi.org/10.1080/00377996.2010.497177; John R. Wunder, ed., *Native American Sovereignty* (Garland Publishing, 1999), v–26; Michael Parrish, "Local Governance and Reform: Local Empowerment," Diné Policy Institute, September 2018; "Navajo Nation Governmental Structure," Navajo Nation Chapters Information, accessed April 16, 2025, https://navajochapters.org/about.

When uranium mines associated: Details about mine waste were synthesized from "The Health and Environmental Impacts of Uranium Contamination in the Navajo Nation: Hearing Before the Committee on Oversight and Government Reform," U.S. Congress, October 23, 2007, https://www.congress.gov/110/chrg/CHRG-110hhrg45611/CHRG-110hhrg45611.pdf; David, "Clean Up Your Act"; "Ten-Year Plan: Federal Actions to Address Impacts of Uranium Contamination on the Navajo Nation," U.S. Environmental Protection Agency, January 2021, https://www.epa.gov/sites/default/files/2021–02/documents/nnaum-ten-year-plan-2021–01.pdf; "Contaminated Structures Program," U.S. Environmental Protection Agency, accessed April 16, 2025, https://www.epa.gov/navajo-nation-uranium-cleanup/structures; Shuey, discussion; Lewis, discussion; Root, discussion; Briehl, discussion.

After advocates pushed for compensation: The law, called the Radiation Exposure Compensation Act, also provided compensation to people who lived downwind of a nuclear testing site in Nevada and had developed certain types of cancers. Details about the law and compensation issues were synthesized from "H.R.2372—Radiation Exposure Compensation Act," U.S. Congress, October 15, 1990, https://www.congress.gov/bill/101st-congress/house-bill/2372/text/pl; Doug Brugge and Rob Goble, "The Radiation Exposure Compensation Act: What Is Fair?," *New Solutions: A Journal of Environmental and Occupational Health Policy* 13, no. 4 (2003): 385–97, https://doi.org/10.2190/N8NK-KB50–5UVV-MXG5; "The Radiation Exposure Compensation Act (RECA): Compensation Related to Exposure to Radiation from Atomic Weapons Testing and Uranium Mining," U.S. Congressional Research Service, July 19, 2024, https://www.congress.gov/crs_external_products/R/PDF/R43956/R43956.15.pdf; Brugge and Goble, "History of Uranium Mining"; Brugge, discussion; Charley, discussion; Goble, discussion.

The act was amended a decade later: The 2000 amendment is described at "Radiation Exposure Compensation Act Amendments of 2000," U.S. Congress, July 10, 2000, https://www.congress.gov/bill/106th-congress/senate-bill/1515/text; "Claims Under the Radiation Exposure Compensation Act Amendments of 2000; Technical Amendments," Federal Register, August 8, 2002, https://www.federalregister.gov/documents/2002/08/07/02–19221/claims-under-the-radiation-exposure-compensation-act-amendments-of-2000-technical-amendments; Brugge and Goble, "Radiation Exposure Compensation Act." The 2025 eligibility expansion is described at "H.R.1—One Big Beautiful Bill Act," U.S. Congress, July 4, 2025, https://www.congress.gov/bill/119th-congress/house-bill/1/text; "Radiation Exposure Compensation Act," U.S. Department of Justice, accessed July 30, 2025, https://www.justice.gov/civil/common/reca; "Hawley Secures Historic RECA Expansion in Base Text of 'Big, Beautiful Bill,'" U.S. Senate, June 12, 2025, https://www.hawley.senate.gov/hawley-secures-historic-reca-expansion-in-base-text-of-big-beautiful-bill; Liam Morris, "A Diamond Provision Hidden in a Rough Bill: Compensation Expanded for Victims of Nuclear Testing," *Utah News Dispatch*, July 28, 2025, https://utahnewsdispatch.com/2025/07/28/a-diamond-provision-hidden-in-a-rough-bill-compensation-expanded-for-victims-of-nuclear-testing.

But to date, no compensation: David, "Clean Up Your Act"; Michelle David (judicial law clerk, U.S. District Court for the Northern District of Georgia), in discussion with the author, November 20, 2023; Yazzie, discussion.

In the late 1990s: Lewis, discussion.

The disease was more severe: Lewis, discussion; Andrew S. Narva, "The Spectrum of Kidney Disease in American Indians," *Kidney International* 63 (February 2003): S3–7, https://doi.org/10.1046/j.1523–1755.63.s83.2.x; Lauren Hund et al., "A Bayesian Framework for Estimating Disease Risk Due to Exposure to Uranium Mine and Mill Waste on the Navajo Nation," *Journal of the Royal Statistical Society Series A: Statistics in Society* 178, no. 4 (October 1, 2015): 1069–91, https://doi.org/10.1111/rssa.12099.

Lewis's team found: Hund et al., "A Bayesian Framework"; Lewis, discussion.

A more recent study: Esther Erdei et al., "Elevated Autoimmunity in Residents Living Near Abandoned Uranium Mine Sites on the Navajo Nation," *Journal of Autoimmunity* 99 (May 2019): 15–23, https://doi.org/10.1016/j.jaut.2019.01.006; Lewis, discussion.

Research on children yielded: Esther Erdei et al., "Environmental Uranium Exposures and Cytokine Profiles Among Mother-Newborn Baby Pairs from the Navajo Birth Cohort Study," *Toxicology and Applied Pharmacology* 456 (December 2022): 116292, https://doi.org/10.1016/j.taap.2022.116292; Lewis, discussion.

On the Navajo Nation, many: Jani C. Ingram et al., "Uranium and Arsenic Unregulated Water Issues on Navajo Lands," *Journal of Vacuum Science & Technology A* 38, no. 3 (May 2020): 031003, https://doi.org/10.1116/1.5142283.

The team started collecting water: Nicole R. Campbell and Jani C. Ingram, "Characterization of 234U/238U Activity Ratios and Potential Inorganic Uranium Complexation Species in Unregulated Water Sources in the Southwest Region of the Navajo Reservation," in *Water Reclamation and Sustainability* (Elsevier, 2014), 77–94, https://doi.org/10.1016/B978-0-12-411645-0.00004–3; Credo et al., "Quantification of Elemental Contaminants"; Ingram, discussion, September 19, 2022; Jonathan Credo (resident in internal medicine and psychiatry, University of California, Davis), in discussion with the author, April 16, 2021.

Finding the wells wasn't easy: Details about wells were synthesized from Ingram, discussion, February 12, 2021, and September 19, 2022; Credo, discussion, April 16, 2021; Shanadeen Begay (adjunct faculty member, Bunker Hill Community College), in discussion with the author, November 3, 2023; Lydia Edgewater (scientist, W. L. Gore & Associates), in discussion with the author, November 20, 2023.

Back at the lab: Details about the water sample tests were synthesized from Credo et al., "Quantification of Elemental Contaminants"; Credo, discussion, April 16, 2021; Edward Pajarillo et al., "Mechanisms of Manganese-Induced Neurotoxicity and the Pursuit of Neurotherapeutic Strategies," *Frontiers in Pharmacology* 13 (December 20, 2022): 1011947, https://doi.org/10.3389/fphar.2022.1011947; Masaki Mogi, "Manganese Exposure Is a Risk for Brain Atrophy," *Hypertension Research* 46, no. 8 (August 2023): 1883–85, https://doi.org/10.1038/s41440-023-01339-2; "Manganese," Royal Society of Chemistry, accessed April 16, 2025, https://periodic-table.rsc.org/element/25/manganese.

Other students in Jani's lab: Details about the sheep research were synthesized from Andee R. Lister, "The Bioaccumulation of Uranium in Sheep Heart and Kidney: The Impact of Contaminated Traditional Food Sources on the Navajo Reservation," (master's diss., Northern Arizona University, 2018), ii–iii, https://openknowledge.nau.edu/id/eprint/5450; Tommy Rock et al., "Traditional Sheep Consumption by Navajo People in Cameron, Arizona," *International Journal of Environmental Research and Public Health* 16, no. 21 (October 30, 2019): 4195, https://doi.org/10.3390/ijerph16214195; Ingram, discussion, May 1, 2017, February 12, 2021, and September 19, 2022; Edgewater, discussion; Andee Lister (PhD student, Northern Arizona University), in discussion with the author, March 18, 2021.

The purpose of the studies: Credo, discussion, April 16, 2021; Colleen Cooley (Diné facilitator and educator), in discussion with the author, February 9, 2023.

it had been discovered in 1789: Marisa J. Monreal and Paula L. Diaconescu, "The Riches of Uranium," *Nature Chemistry* 2, no. 5 (May 2010): 424, https://doi.org/10.1038/nchem.642.

The element was present: Grenthe et al., "Uranium," 271–73, 297–98; Eichstaedt, *If You Poison Us*, 11.

In the late 1930s: Peter Brix and Carsten Salander, "The Discovery of Uranium Fission: Its Intricate History and Far-Reaching Consequences," *Interdisciplinary Science Reviews* 15, no. 4 (December 1990): 314–26, https://doi.org/10.1179/isr.1990.15.4.314; "Otto Hahn: Facts," Nobel Prize, accessed April 17, 2025, https://www.nobelprize.org/prizes/chemistry/1944/hahn/facts; "Otto Hahn: Biographical," Nobel Prize, accessed April 17, 2025, https://www.nobelprize.org/prizes/chemistry/1944/hahn/biographical; Otto Hahn, "From the Natural Transmutations of Uranium to Its Artificial Fission," Nobel Prize, December 13, 1946, https://www.nobelprize.org/uploads/2018/06/hahn-lecture.pdf.

If neutrons hit: Lise Meitner, "Disintegration of Uranium by Neutrons: A New

Type of Nuclear Reaction," *Nature*, 1939; Tom Barton, "Ames Laboratory and Uranium Production in World War II," American Chemical Society, 2022, https://www.acs.org/content/dam/acsorg/education/whatischemistry/landmarks/ames-uranium-production/ames-lab-booklet.pdf; Goble, discussion.

Lise Meitner, an Austrian physicist: Brix and Salander, "Discovery of Uranium Fission."

Government workers were trying to: The lay term "clean up" is often used to describe the process, but Shuey cautions that no mine site can ever be fully rid of all toxic material. The more technical term, "remediation," refers to tasks such as covering or burying waste, or moving it to a disposal facility, in hopes of reducing the chances that people will come into direct contact or that contaminants will spread to other places. "Ten-Year Plan: Federal Actions"; Shuey, discussion; Brugge, discussion.

In 2007, Congressman Henry Waxman: "Health and Environmental Impacts"; David, discussion.

Under the so-called Superfund: David, "Clean Up Your Act"; David, discussion; Shuey, discussion; "Superfund: CERCLA Overview," U.S. Environmental Protection Agency, accessed April 17, 2025, https://www.epa.gov/superfund/superfund-cercla-overview.

As of this writing, agreements: "Abandoned Mines Cleanup"; "Abandoned Uranium Mine Settlements On or Near the Navajo Nation," U.S. Environmental Protection Agency, November 2022, https://www.epa.gov/system/files/documents/2023–02/epa-factsheet-abandoned-uranium-mine-settlements-on-the-navajo-nation.pdf; "Ten-Year Plan: Federal Actions"; Will Duncan (assistant director in Superfund and emergency management division, EPA), in discussion with the author, June 11, 2025.

progress felt frustratingly slow: Shuey, discussion; Eric Jantz (legal director, New Mexico Environmental Law Center), in discussion with the author, June 6, 2025.

settlements would address roughly two-thirds: This estimate refers to the fraction of waste, not the fraction of sites. Duncan estimates that there are about sixteen million tons of waste rock on the Navajo Nation, and the settlements would account for more than eleven million tons.

the EPA announced a plan: "EPA and Navajo Nation Select Cleanup Plan to Remove Waste at Quivira Mines Site," U.S. Environmental Protection Agency, January 7, 2025, https://www.epa.gov/newsreleases/epa-and-navajo-nation-select-cleanup-plan-remove-waste-quivira-mines-site.

Many mines were "orphaned": Details about orphan mines and the federal government's role in mining were synthesized from David, "Clean Up Your Act"; David, discussion; Charley, discussion.

federal government is paying for cleanup: Duncan, discussion; "Abandoned

Uranium Mine Settlements"; "Trust Mines Background and Site Updates," Environmental Protection Agency, July 2020, https://www.epa.gov/sites/default/files/2021–01/documents/nnaum-trust-mines-factsheet-2020–07.pdf; "Trust Mines: Legal Documents and Settlements," Environmental Protection Agency, accessed July 24, 2025, https://www.epa.gov/navajo-nation-uranium-cleanup/trust-mines-legal-documents.

The government channeled the equivalent: Figures vary, but the cost is typically estimated at around two billion dollars, which I extrapolated to today's dollars based on inflation since the 1940s. See "Manhattan Project Background Information and Preservation Work," U.S. Department of Energy, accessed April 17, 2025, https://www.energy.gov/lm/manhattan-project-background-information-and-preservation-work; Catie Edmondson, "A Reporter's Journey into How the U.S. Funded the Bomb," *New York Times*, updated January 18, 2024, https://www.nytimes.com/2024/01/17/us/politics/atomic-bomb-secret-funding-congress.html; "The Manhattan Project: The Race to Build the Atomic Bomb," *Bulletin of the Atomic Scientists*, accessed April 17, 2025, https://thebulletin.org/virtual-tour/the-manhattan-project-the-race-to-build-the-atomic-bomb.

NuFuels, a subsidiary: The company has previously gone by the name Hydro Resources, Inc. Details about the company and proposed mining were synthesized from "Crownpoint-Churchrock Uranium Project," Laramide Resources, accessed April 17, 2025, https://laramide.com/projects/crownpoint-churchrock-uranium-project; "Technical Report on the Churchrock Uranium Project, McKinley County, New Mexico, USA," Laramide Resources, January 8, 2024, https://wp-laramide-2023.s3.ca-central-1.amazonaws.com/media/2024/01/SLR-Laramide-Churchrock-NI-43–101-Report-FINAL-Jan-8–20.pdf; "Laramide Resources Ltd. Announces the Engagement of SLR International Corporation to Produce a Preliminary Economic Assessment on the Crownpoint/Churchrock Uranium Project," Laramide Resources, January 23, 2023, https://laramide.com/laramide-resources-ltd-announces-the-engagement-of-slr-international-corporation-to-produce-a-preliminary-economic-assessment-on-the-crownpoint-churchrock-uranium-project; "Proposed Crownpoint Uranium Mine," New Mexico Environmental Law Center, accessed April 17, 2025, https://nmelc.org/our-work/cases/crownpoint-proposed-uranium-mine; "Petition by Eastern Navajo Diné Against Uranium Mining and Mitchell Capitan, Rita Capitan, Christine Smith, Keithlynn Smith, Kenneth Smith, and Larry King on Their Own Behalf Against the United States of America," New Mexico Environmental Law Center, May 13, 2011, https://nmelc.org/wp-content/uploads/2021/07/endaum_final_petition_with_figures-1.pdf; letter to Jonathan Perry, U.S. Nuclear Regulatory Commission, October

26, 2022, https://www.nrc.gov/docs/ML2227/ML22270A090.pdf; Jantz, in discussion with the author, November 29, 2023.

The proposed sites: "Petition by Eastern Navajo Diné"; Hannah Grover, "Navajo Nation Officials, Activists Feel Cut Out as Company Advances Uranium Mining Plans," *New Mexico Political Report*, May 1, 2023, https://nmpoliticalreport.com/2023/05/01/navajo-nation-officials-activists-feel-cut-out-as-company-advances-uranium-mining-plans.

In 2023, Laramide announced: "Laramide Resources Ltd. Announces Results from the Diamond Drilling Program at its Crownpoint-Churchrock Uranium Project, New Mexico, U.S.A.," Laramide Resources, March 24, 2023, https://laramide.com/laramide-resources-ltd-announces-results-from-the-diamond-drilling-program-at-its-crownpoint-churchrock-uranium-project-new-mexico-u-s-a.

the following year, the company: "Laramide PEA Confirms Potential for Long-Life, Low-Cost In Situ Recovery Uranium Operation at Churchrock Project, New Mexico, USA," Laramide Resources, January 11, 2024, https://laramide.com/laramide-pea-confirms-potential-for-long-life-low-cost-in-situ-recovery-uranium-operation-at-churchrock-project-new-mexico-usa.

The firm planned to use: "Technical Report on the Churchrock"; "Petition by Eastern Navajo Diné"; Jantz, discussion.

Jantz was skeptical of claims: "Technical Report on the Churchrock"; "Laramide PEA Confirms Potential."

After challenging the license: "Petition by Eastern Navajo Diné"; Jantz, discussion.

Paleontology: Through a Glass Darkly

Details about the *Lystrosaurus* tusk research were drawn from interviews with team members Meg Whitney or Chris Sidor; from my reporting trips to Whitney's lab at the University of Washington in Seattle; or from the scientific paper Megan R. Whitney and Christian A. Sidor, "Evidence of Torpor in the Tusks of *Lystrosaurus* from the Early Triassic of Antarctica," *Communications Biology* 3, no. 1 (August 27, 2020): 471, https://doi.org/10.1038/s42003-020-01207-6. Biographical details about Whitney were drawn from interviews with Whitney, her parents Peter and Leslie Whitney, and her former advisor Kristi Curry Rogers. Information about *Lystrosaurus* and paleontology background were drawn from interviews with team members or experts such as Ken Angielczyk, Jennifer Botha, Christian Kammerer, Brandon Peecook, and Bill Hammer. Information about fieldwork in Antarctica was drawn from interviews with Whitney, Sidor, Peecook, and Hammer.

A stumpy, four-legged animal: The information in this paragraph was synthesized from Ken Angielczyk (curator of paleomammalogy, Field Museum), in discussion with the author, March 6, 2019; Jennifer Botha (director of GENUS, University of the Witwatersrand), in discussion with the author, August 12, 2019; Meg Whitney (vertebrate paleontologist, Loyola University Chicago), in discussion with the author, November 28, 2018; Edwin H. Colbert, *Wandering Lands and Animals: The Story of Continental Drift and Animal Populations* (New York: Dover Publications, 1985), 26–27; Jennifer Botha and Roger M. H. Smith, "*Lystrosaurus* Species Composition Across the Permo-Triassic Boundary in the Karoo Basin of South Africa," *Lethaia* 40, no. 2 (April 8, 2007): 125–37, https://doi.org/10.1111/j.1502–3931.2007.00011.x; J. Botha and R. M. H. Smith, "Biostratigraphy of the *Lystrosaurus declivis* Assemblage Zone (Beaufort Group, Karoo Supergroup), South Africa," *South African Journal of Geology* 123, no. 2 (June 1, 2020): 207–16, https://doi.org/10.25131/sajg.123.0015; "Frequently Asked Questions: Torpor in Antarctic *Lystrosaurus*," *UW News*, August 27, 2020, https://www.washington.edu/news/2020/08/27/faq-torpor-lystrosaurus.

tusks grew layers: Whitney and Sidor, "Evidence of Torpor."

These animals thrived: *Lystrosaurus* is a genus (a group of related species), not just one species. While some individual species of *Lystrosaurus* didn't survive for long after the mass extinction event, the genus did. Angielczyk, discussion, March 6, 2019; Botha and Smith, "*Lystrosaurus* Species Composition"; Jennifer Botha, "The Paleobiology and Paleoecology of South African *Lystrosaurus*," *PeerJ* 8 (November 24, 2020): e10408, https://doi.org/10.7717/peerj.10408; Pia A. Viglietti et al., "Evidence from South Africa for a Protracted End-Permian Extinction on Land," *Proceedings of the National Academy of Sciences* 118, no. 17 (April 27, 2021): e2017045118, https://doi.org/10.1073/pnas.2017045118.

Around 252 million years ago: Information about the chain of events leading to the mass extinction was synthesized from Viglietti et al., "Evidence from South Africa"; S. D. Burgess, J. D. Muirhead, and S. A. Bowring, "Initial Pulse of Siberian Traps Sills as the Trigger of the End-Permian Mass Extinction," *Nature Communications* 8, no. 1 (July 31, 2017): 164, https://doi.org/10.1038/s41467-017-00083-9; Ryosuke Saito et al., "Centennial Scale Sequences of Environmental Deterioration Preceded the End-Permian Mass Extinction," *Nature Communications* 14, no. 1 (April 14, 2023): 2113, https://doi.org/10.1038/s41467-023-37717-0; Michael J. Benton, "Hyperthermal-Driven Mass Extinctions: Killing Models During the Permian–Triassic Mass Extinction," *Philosophical Transactions of the Royal Society A: Mathematical, Physical, and Engineering Sciences* 376, no. 2130 (October 13, 2018): 20170076, https://doi.org/10.1098/rsta.2017.0076; Lee Kump, "Climate Change and Marine Mass

Extinction," *Science* 362, no. 6419 (December 7, 2018): 1113–14, https://doi.org/10.1126/science.aav736; K. M. Meyer, L. R. Kump, and A. Ridgwell, "Biogeochemical Controls on Photic-Zone Euxinia during the End-Permian Mass Extinction," *Geology* 36, no. 9 (2008): 747, https://doi.org/10.1130/G24618A.1; Martin Schobben et al., "A Nutrient Control on Marine Anoxia During the End-Permian Mass Extinction," *Nature Geoscience* 13, no. 9 (September 2020): 640–46, https://doi.org/10.1038/s41561-020-0622-1; Angielczyk, discussion, March 6, 2019; Botha, discussion, August 12, 2019; Brandon Peecook (paleobiologist, Idaho State University), in discussion with the author, March 7, 2019; Peecook, email message to author, June 11, 2025.

81 to 96 percent: The 81 percent figure for marine species is from Steven M. Stanley, "Estimates of the Magnitudes of Major Marine Mass Extinctions in Earth History," *Proceedings of the National Academy of Sciences* 113, no. 42 (October 18, 2016), https://doi.org/10.1073/pnas.1613094113. A higher estimate of up to 96 percent for marine species is from David M. Raup, "Size of the Permo-Triassic Bottleneck and Its Evolutionary Implications," *Science* 206, no. 4415 (1979): 217–18, https://doi.org/10.1126/science.206.4415.217. An estimated 89 percent of tetrapod genera (groups of species) went extinct, according to Michael J. Benton et al., "The First Half of Tetrapod Evolution, Sampling Proxies, and Fossil Record Quality," *Palaeogeography, Palaeoclimatology, Palaeoecology* 372 (February 2013): 18–41, https://doi.org/10.1016/j.palaeo.2012.09.005. Experts' estimates for tetrapod species extinctions range from around 89 to 96 percent; Angielczyk, email message to author, June 27, 2025, and Michael Benton (vertebrate paleontologist, University of Bristol), email message to author, July 8, 2025. Note that the estimate for extinctions of entire tetrapod families (groups of genera) is lower at around two-thirds; see Sarda Sahney and Michael J. Benton, "Recovery from the Most Profound Mass Extinction of All Time," *Proceedings of the Royal Society B: Biological Sciences* 275, no. 1636 (April 7, 2008): 759–65, https://doi.org/10.1098/rspb.2007.1370.

But somehow, *Lystrosaurus*: Botha and Smith, "*Lystrosaurus* Species Composition."

Perhaps creatures with similar traits: Angielczyk, discussion, March 6, 2019.

Neither could I: Botha, discussion, August 12, 2019; Angielczyk, discussion, March 6, 2019; Bill Hammer (paleontologist, Augustana College), in discussion with the author, March 6, 2019.

Some benefits are relatively modest: Whitney, in discussion with the author, July 24, 2019.

Mammals' ability to lactate: Whitney, discussion, July 24, 2019; Michael J. Benton, "The Origin of Endothermy in Synapsids and Archosaurs and Arms

Races in the Triassic," *Gondwana Research* 100 (December 2021): 261–89, https://doi.org/10.1016/j.gr.2020.08.003.

***Lystrosaurus* was one of those:** Christian Kammerer (paleontologist, North Carolina Museum of Natural Sciences), in discussion with the author, September 12, 2019.

It belonged to a group: Botha, "Paleobiology and Paleoecology"; Benton, "Origin of Endothermy"; Angielczyk, discussion, March 6, 2019; Kammerer, discussion, September 12, 2019.

Some therapsids eventually evolved: Jennifer Botha and Adam Huttenlocker, "Nonmammalian Synapsids," in *Vertebrate Skeletal Histology and Paleohistology*, by Vivian De Buffrénil et al., 1st ed. (Boca Raton: CRC Press, 2021), 550–63, https://doi.org/10.1201/9781351189590–28; Kammerer, discussion, September 12, 2019.

They ranged from: Angielczyk, discussion, March 6, 2019.

Mole-like creatures: Kenneth D. Angielczyk and Christian F. Kammerer, "Non-Mammalian Synapsids: The Deep Roots of the Mammalian Family Tree," in *Mammalian Evolution, Diversity, and Systematics*, eds. Frank Zachos and Robert Asher (De Gruyter, 2018), 154, https://doi.org/10.1515/9783110341553–005; Angielczyk, discussion, March 6, 2019; Kammerer, discussion, September 12, 2019.

In the subgroup that evolved: Kammerer, discussion, September 12, 2019; Frederick Tolchard et al., "A New Large Gomphodont from the Triassic of South Africa and Its Implications for Gondwanan Biostratigraphy," *Journal of Vertebrate Paleontology* 41, no. 2 (March 4, 2021): e1929265, https://doi.org/10.1080/02724634.2021.1929265; Jackie Adams, "Wrestling with Rats and Other Fascinating Stories About The Princess Bride," *Los Angeles Magazine*, October 13, 2014, https://lamag.com/film/wrestling-rubber-rats-fascinating-stories-princess-bride.

The second advantage: The information in this paragraph was synthesized from Zoe T. Kulik et al., "Living Fast in the Triassic: New Data on Life History in *Lystrosaurus* (Therapsida: Dicynodontia) from Northeastern Pangea," *PLOS One* 16, no. 11 (November 5, 2021): e0259369, https://doi.org/10.1371/journal.pone.0259369; Megan R. Whitney, "Histological Insights into Trait Acquisition in Non-Mammalian Synapsids" (PhD diss., University of Washington, 2019), 61, https://digital.lib.washington.edu/researchworks/items/38f73608-b8d9–4ebf-929b-a92464342675; Whitney, discussion, November 28, 2018.

The tusk that Meg: The information in this paragraph was synthesized from Whitney, discussion, January 16, 2019; Whitney and Sidor, "Evidence of Torpor"; Hammer, discussion.

Antarctica and South Africa, which had: D. H. Elliot et al., "Triassic

Tetrapods from Antarctica: Evidence for Continental Drift," *Science* 169, no. 3951 (September 18, 1970): 1197–1201, https://doi.org/10.1126/science.169.3951.1197.

Though Antarctica was far warmer: Chris Sidor (paleontologist, University of Washington), in discussion with the author, March 7, 2019; Peecook, discussion.

In the past, scientists had: The information in this paragraph was synthesized from Kammerer, discussion, September 12, 2019, and September 17, 2020; Botha, discussion, August 12, 2019, and September 29, 2020; Whitney, discussion, July 24, 2019, and March 11, 2024; Jennifer Botha-Brink and Kenneth D. Angielczyk, "Do Extraordinarily High Growth Rates in Permo-Triassic Dicynodonts (Therapsida, Anomodontia) Explain Their Success Before and After the End-Permian Extinction?," *Zoological Journal of the Linnean Society* 160, no. 2 (July 26, 2010): 341–65, https://doi.org/10.1111/j.1096-3642.2009.00601.x; Benton, "Origin of Endothermy."

mammals have muscular diaphragms: The information in this paragraph was synthesized from Botha, discussion, August 12, 2019, and September 29, 2020; Benton, "Origin of Endothermy."

Some animals fell along: Kammerer, discussion, September 12, 2019.

swordfish warm up their brains: Kerstin A. Fritsches, Richard W. Brill, and Eric J. Warrant, "Warm Eyes Provide Superior Vision in Swordfishes," *Current Biology* 15, no. 1 (January 2005): 55–58, https://doi.org/10.1016/j.cub.2004.12.064.

Leatherback sea turtles: Brian L. Bostrom et al., "Behaviour and Physiology: The Thermal Strategy of Leatherback Turtles," ed. Lewis George Halsey, *PLOS One* 5, no. 11 (November 10, 2010): e13925, https://doi.org/10.1371/journal.pone.0013925.

eerily glowing bacteria: Athina Rodou, Dennis O. Ankrah, and Christos Stathopoulos, "Toxins and Secretion Systems of *Photorhabdus luminescens*," *Toxins* 2, no. 6 (June 1, 2010): 1250–64, https://doi.org/10.3390/toxins2061250.

a pill bug–like creature: Richard C. Brusca and Matthew R. Gilligan, "Tongue Replacement in a Marine Fish (*Lutjanus guttatus*) by a Parasitic Isopod (Crustacea: Isopoda)," *Copeia* 1983, no. 3 (August 16, 1983): 813, https://doi.org/10.2307/1444352.

it was likely a hatching: Kristina Curry Rogers et al., "Precocity in a Tiny Titanosaur from the Cretaceous of Madagascar," *Science* 352, no. 6284 (April 22, 2016): 450–53, https://doi.org/10.1126/science.aaf1509.

the continent is covered: "Ice Sheets," U.S. National Science Foundation, accessed February 20, 2024, https://www.nsf.gov/geo/opp/antarct/science/icesheet.jsp.

It often lived on: The information in this paragraph was synthesized from

Angielczyk, discussion, March 6, 2019; Botha, discussion, August 12, 2019; Kammerer, discussion, September 12, 2019; Botha and Huttenlocker, "Nonmammalian Synapsids."

When the Siberian volcano: Angielczyk, discussion, March 6, 2019; Whitney, discussion, November 28, 2018.

maybe they were already accustomed: Kammerer, discussion, September 12, 2019.

On a typical day: Kammerer, discussion, September 12, 2019.

"Permian" and "Triassic": "International Chronostratigraphic Chart," International Commission on Stratigraphy, accessed February 19, 2025, https://stratigraphy.org/timescale.

Journalist John McPhee described: John McPhee, *Basin and Range* (New York: Farrar, Straus and Giroux, 1981), 135.

Seattle area was hit: Susan Miller, "Seattle Is Having Its Snowiest February in 70 Years—and More Is on the Way," *USA Today*, February 10, 2019, https://www.usatoday.com/story/news/nation/2019/02/10/seattle-snowiest-february-70-years-more-coming/2830164002.

In his book *On the Origin*: Charles Darwin, *On the Origin of Species* (London: John Murray, 1859), 310–11.

Other scientists expressed caution: The caveats in this paragraph are from Botha, discussion, September 29, 2020; Kammerer, discussion, September 17, 2020.

Ken Angielczyk and an international: The information about the ear-canal study was synthesized from Ricardo Araújo et al., "Inner Ear Biomechanics Reveals a Late Triassic Origin for Mammalian Endothermy," *Nature* 607, no. 7920 (July 28, 2022): 726–31, https://doi.org/10.1038/s41586-022-04963-z; Stefan Glasauer and Hans Straka, "Ear Pins Down Evolution of Thermoregulation," *Nature*, July 28, 2022, 661–62; Julien Benoit et al., "Mystery Solved: When Mammals' Ancestors Became Warm-Blooded," *The Conversation*, July 20, 2022, https://theconversation.com/mystery-solved-when-mammals-ancestors-became-warm-blooded-186359; Angielczyk, in discussion with the author, February 28, 2024; Botha and Smith, "Biostratigraphy of the *Lystrosaurus declivis*"; "Frequently Asked Questions: Torpor."

It brought to mind: Colbert, *Wandering Lands*, 19.

Epilogue: An Act of Humility

In 1929, German meteorologist: The story of Wegener's expedition and death was synthesized from Cornelia Lüdecke, "Lifting the Veil: The Circumstances That Caused Alfred Wegener's Death on the Greenland Icecap, 1930,"

Polar Record 36, no. 197 (April 2000): 139–54, https://doi.org/10.1017/S0032247400016247; J. Abermann et al., "Learning from Alfred Wegener's Pioneering Field Observations in West Greenland after a Century of Climate Change," *Scientific Reports* 13, no. 1 (May 23, 2023): 7583, https://doi.org/10.1038/s41598-023-33225-9; "Death on the Eternal Ice," Alfred Wegener Institute, October 21, 2005, https://www.awi.de/en/about-us/service/press/archive/archive-single-view/death-on-the-eternal-ice.html; Roger M. McCoy, *Ending in Ice: The Revolutionary Idea and Tragic Expedition of Alfred Wegener* (Oxford University Press, 2006), 101.

About two decades earlier: Details about Wegener's proposal were synthesized from Alfred Wegener, "The Origins of Continents" (lecture at a general meeting of the Geologische Vereinigung, Frankfurt, January 6, 1912), trans. Roland von Huene, *International Journal of Earth Sciences* 91 (2002): S4–S17, https://doi.org/10.1007/s00531-002-0271-1; Naomi Oreskes, *Plate Tectonics: An Insider's History of the Modern Theory of the Earth* (Westview Press, 2001), 3, 7; Naomi Oreskes, "Building Scientific Knowledge: The Story of Plate Tectonics," *HHMI Biointeractive*, accessed April 24, 2025, https://www.biointeractive.org/professional-learning/science-talks/building-scientific-knowledge-story-plate-tectonics; Naomi Oreskes, "How Plate Tectonics Clicked," *Nature* 501 (September 5, 2013): 27–29, https://doi.org/10.1038/501027a; James Powell, *Four Revolutions in the Earth Sciences: From Heresy to Truth* (Columbia University Press, 2015), 65; "The Copernicus of Geosciences: Alfred Wegener Presented His Revolutionary Theory of Continental Drift 100 Years Ago," Alfred Wegener Institute, December 21, 2011, https://www.awi.de/en/about-us/service/press/archive/archive-single-view/der-kopernikus-der-geowissenschaften-vor-fast-100-jahren-verkuendete-alfred-wegener-seine-revolutionaere-theorie-von-der-verschiebung-der-kontinente.html; Jon Powell (geologist, Fort Lewis College), in discussion with the author, August 9, 2019; Hannah Davies (geophysicist, University of Lisbon), in discussion with the author, September 10, 2019. Details about earlier scholars' remarks were drawn from James Romm, "A New Forerunner for Continental Drift," *Nature* 367 (February 3, 1994): 407–8, https://doi.org/10.1038/367407a0; the theologian's quote appears on p. 407.

Other researchers had come up: Naomi Oreskes, *The Rejection of Continental Drift: Theory and Method in American Earth Science* (Oxford University Press, 1999), 11; Oreskes, *Plate Tectonics*, 3–5; Wegener, "The Origins of Continents."

In the 1800s, scientists believed: Oreskes, *Plate Tectonics*, 4; Powell, *Four Revolutions*, 66–67.

But new findings had started: Oreskes, *Plate Tectonics*, 6–7; Powell, *Four Revolutions*, 67–68; Wegener, "The Origins of Continents."

In 1896, French physicist: "Henri Becquerel: Facts," Nobel Prize, accessed April 24, 2025, https://www.nobelprize.org/prizes/physics/1903/becquerel/facts; Henri Becquerel, "On Radioactivity, A New Property of Matter," Nobel Prize, December 11, 1903, https://www.nobelprize.org/prizes/physics/1903/becquerel/lecture; "Marie Curie: Facts," Nobel Prize, accessed April 24, 2025, https://www.nobelprize.org/prizes/physics/1903/marie-curie/facts; Pierre Curie, "Radioactive Substances, Especially Radium," Nobel Prize, June 6, 1905, https://www.nobelprize.org/uploads/2018/06/pierre-curie-lecture.pdf; Oreskes, *Rejection of Continental Drift*, 48.

the "decay" of radioactive elements: It turns out that scientists weren't entirely wrong; Earth almost certainly has been cooling and contracting. But at the time, the new findings about radioactivity threw those ideas into doubt. Becquerel, "On Radioactivity"; Curie, "Radioactive Substances"; Oreskes, *Rejection of Continental Drift*, 50; Oreskes, *Plate Tectonics*, 7; Powell, *Four Revolutions*, 68; Wegener, "The Origins of Continents"; John Joly, "Uranium and Geology," *Science* 28, no. 725 (November 20, 1908): 697–713, https://doi.org/10.1126/science.28.725.697; Patrick N. Wyse Jackson, "John Joly's Paper: Uranium and Geology (1908)," *Episodes* 25, no. 4 (December 1, 2002): 258–63, https://doi.org/10.18814/epiiugs/2002/v25i4/007; Robert John Strutt, "On the Distribution of Radium in the Earth's Crust and on the Earth's Internal Heat," *Proceedings of the Royal Society A* 77 (1906): 472–85, http://doi.org/10.1098/rspa.1906.0042; Oreskes, *Rejection of Continental Drift*, 50; Brad Foley (mantle dynamicist, Pennsylvania State University), in discussion with the author, June 9, 2025.

"Are we living": Joly, "Uranium and Geology," 705, 706.

Meanwhile, other scientists found folds: Oreskes, *Plate Tectonics*, 6; Powell, *Four Revolutions*, 67.

Still, that didn't mean that geologists: Alfred Wegener, *The Origin of Continents and Oceans*, 4th ed., trans. John Biram (1929; repr., Dover Publications, 1966), 17.

Though he had scattered supporters: Oreskes, *Plate Tectonics*, 7–12; Oreskes, "Building Scientific Knowledge"; Powell, *Four Revolutions*, 87–92; Colbert, *Wandering Lands*, 12–13; H. H. Hess, "History of Ocean Basins," in *Petrologic Studies*, eds. A. E. J. Engel, Harold L. James, and B. F. Leonard (USA: Geological Society of America, 1962), 599–620, https://doi.org/10.1130/Petrologic.1962.599. The skeptics' quotes appear in Powell, *Four Revolutions*, 89, 91.

In some circles, Wegener continued: Colbert, *Wandering Lands*, 12–13; Oreskes, *Plate Tectonics*, xxiv.

And yet evidence eventually emerged: Oreskes, *Plate Tectonics*, xi; Oreskes, "Building Scientific Knowledge"; Oreskes, "How Plate Tectonics Clicked."

From the 1930s to 1950s: Oreskes, *Plate Tectonics*, 13–23; Oreskes, "How Plate Tectonics Clicked"; Powell, *Four Revolutions*, 109–12; John Milsom, "Magnetic Stripes, Ron Mason, and the RAS," *Astronomy & Geophysics* 59, no. 2 (April 2018): 2.23–2.25, https://doi.org/10.1093/astrogeo/aty081; Suzanne OConnell, "Marie Tharp Pioneered Mapping the Bottom of the Ocean 6 Decades Ago—Scientists Are Still Learning About Earth's Last Frontier," *The Conversation*, July 28, 2020, https://theconversation.com/marie-tharp-pioneered-mapping-the-bottom-of-the-ocean-6-decades-ago-scientists-are-still-learning-about-earths-last-frontier-142451.

These findings could be explained: The explanation of plate tectonics was synthesized from Davies, discussion; Foley, discussion; Christopher Spencer (geologist, Queen's University), in discussion with the author, June 23, 2025; Mark Behn (geodynamicist, Boston College), in discussion with the author, July 8, 2025; Powell, discussion; David Elliot (geologist, The Ohio State University), in discussion with the author, August 12, 2019; Oreskes, "How Plate Tectonics Clicked"; Oreskes, "Building Scientific Knowledge"; Powell, *Four Revolutions*, 152–56; Arthur Holmes, "Radioactivity and Earth Movements," *Nature* 128, no. 496 (1931): 559–606, https://doi.org/10.1038/128496e0; Robert S. Dietz, "Continent and Ocean Basin Evolution by Spreading of the Sea Floor," *Nature* 190, no. 4779 (June 1961): 854–57, https://doi.org/10.1038/190854a0; Carolyn Gramling, "Birth of a Theory," *Science News*, accessed April 24, 2025, https://www.sciencenews.org/century/earth-history-plate-tectonics-volcanoes-earthquakes; Walter Sullivan, "The Restless Continents," *New York Times*, January 12, 1975, https://www.nytimes.com/1975/01/12/archives/the-restless-continents.html; OConnell, "Marie Tharp Pioneered Mapping"; KamLAND Collaboration, "Partial Radiogenic Heat Model for Earth Revealed by Geoneutrino Measurements," *Nature Geoscience* 4, no. 9 (September 2011): 647–51, https://doi.org/10.1038/ngeo1205.

In the mid-1800s: Alfred Russel Wallace, *The Malay Archipelago: The Land of the Orang-utan, and the Bird of Paradise. A Narrative of Travel, with Studies of Man and Nature* (London: Macmillan and Co., 1869), 1: x–xii, https://wallace-online.org/converted/pdf/1869_MalayArchipelago_S715.1.pdf; Alfred Russel Wallace, "On the Physical Geography of the Malay Archipelago," *Journal of the Royal Geographical Society of London* 33 (1863): 217–34, https://www.jstor.org/stable/1798448; James McNish, "Who Was Alfred Russel Wallace?," Natural History Museum, accessed April 24, 2025, https://www.nhm.ac.uk/discover/who-was-alfred-russel-wallace.html; Arved Kirschbaum, "Wallace: the Evolution Theorist You May Not Know

About," Kew Gardens, February 13, 2020, https://www.kew.org/read-and-watch/the-wallace.

On Borneo, Sumatra, Java: Wallace, "On the Physical Geography"; Kirschbaum, "Wallace: the Evolution Theorist"; "Dividing Species: Wallace Line Map," *National Geographic*, accessed April 24, 2025, https://education.nationalgeographic.org/resource/dividing-species-wallace-line-map. Wallace's quote appears in Wallace, "On the Physical Geography," 232.

With plate tectonics, scientists could: The explanation of plate tectonics and the Wallace Line was synthesized from Larry Heaney (curator of mammals, Field Museum), in discussion with the author, July 3, 2025; Spencer, discussion; Jason R. Ali and Lawrence R. Heaney, "Wallace's Line, Wallacea, and Associated Divides and Areas: History of a Tortuous Tangle of Ideas and Labels," *Biological Reviews* 96, no. 3 (June 2021): 922–42, https://doi.org/10.1111/brv.12683; A. Skeels et al., "Paleoenvironments Shaped the Exchange of Terrestrial Vertebrates Across Wallace's Line," *Science* 381, no. 6653 (July 7, 2023): 86–92, https://doi.org/10.1126/science.adf7122; "The Invisible Barrier Keeping Two Worlds Apart," PBS, May 2, 2023, https://www.pbs.org/video/wallace-line-xgpfxa.

By the mid-1960s: Oreskes, "How Plate Tectonics Clicked"; Oreskes, *Plate Tectonics*, xxiii, 25; Colbert, *Wandering Lands*, xvi; Elliot, discussion; Powell, discussion.

Then in 1967, a geologist: Peter J. Barrett, Ralph J. Baillie, and Edwin H. Colbert, "Triassic Amphibian from Antarctica," *Science* 161, no. 3840 (August 2, 1968): 460–62, doi:10.1126/science.161.3840.460; "Drilling into Peter Barrett's Career," Victoria University of Wellington, February 26, 2021, https://www.wgtn.ac.nz/antarctic/about/news/drilling-into-peter-barretts-career; D. H. Elliot et al., "Triassic Tetrapods from Antarctica: Evidence for Continental Drift," *Science* 169, no. 3951 (September 18, 1970): 1197–1201, https://doi.org/10.1126/science.169.3951.1197; Colbert, *Wandering Lands*, xvi, 15–16; Elliot, discussion.

Still, one could nitpick: Elliot, discussion; Powell, discussion.

So Edwin Colbert, a paleontologist: Elliot, discussion; Powell, discussion; Colbert, *Wandering Lands*, 16; Elliot et al., "Triassic Tetrapods from Antarctica."

After about a week: Colbert, *Wandering Lands*, 25.

Many fossils of these creatures: Colbert, *Wandering Lands*, 25–28.

Shortly afterward, two team members: Colbert, *Wandering Lands*, 18, 31–32; Elliot et al., "Triassic Tetrapods from Antarctica"; Powell, discussion. Colbert's quote appears in Colbert, *Wandering Lands*, 32.

The discovery was "indisputable" evidence: Colbert, *Wandering Lands*, 32.

There was no way: Walter Sullivan, "Fossil Called Proof of Continent Link," *New York Times*, December 6, 1969, https://www.nytimes.com/1969/12/06/archives/fossil-called-proof-of-continent-link-fossil-called-proof-that-all.html.

After the rise of *Lystrosaurus*: "Frequently Asked Questions: Torpor"; Sam Rae and Lisa Hendry, "What Killed the Dinosaurs?," Natural History Museum, accessed April 24, 2025, https://www.nhm.ac.uk/discover/dinosaur-extinction.html; Benton, "Origin of Endothermy"; Benoit et al., "Mystery Solved"; Philip Hunter, "The Rise of the Mammals: Fossil Discoveries Combined with Dating Advances Give Insight into the Great Mammal Expansion," *EMBO Reports* 21, no. 11 (November 5, 2020): e51617, https://doi.org/10.15252/embr.202051617; Steve Brusatte, "How Mammals Came to Dominate the World," *Science Friday*, June 17, 2022, https://www.sciencefriday.com/articles/rise-and-reign-of-mammals-book.

We domesticated billions: Bill Chappell, "Along with Humans, Who Else Is in the 7 Billion Club?," NPR, November 3, 2011, https://www.npr.org/sections/thetwo-way/2011/11/03/141946751/along-with-humans-who-else-is-in-the-7-billion-club.

We drove other species: Gerardo Ceballos and Paul R. Ehrlich, "Mutilation of the Tree of Life via Mass Extinction of Animal Genera," *Proceedings of the National Academy of Sciences* 120, no. 39 (September 26, 2023): e2306987120, https://doi.org/10.1073/pnas.2306987120.

We burned so much: Lee and Romero, eds., "Summary for Policymakers," 4–5.

We harnessed the power: Brix and Salander, "Discovery of Uranium Fission"; "Otto Hahn: Biographical"; Hahn, "From the Natural Transmutations"; "Ten-Year Plan: Federal Actions"; Shuey, discussion.

As we claimed more habitat: Laura Tangley, "In Harm's Way," National Wildlife Federation, February 1, 2021, https://www.nwf.org/Home/Magazines/National-Wildlife/2021/Feb-Mar/Conservation/Zoonotic-Diseases; "How Wildlife Exploitation and Habitat Loss Fuel Pandemic Risk," Center for Biological Diversity, accessed April 24, 2025, https://www.biologicaldiversity.org/campaigns/wildlife-exploitation-and-pandemic-risk/index.html; Emma Gosalvez, "How Habitat Destruction Enables the Spread of Diseases Like COVID-19," NC State University, April 22, 2020, https://cnr.ncsu.edu/news/2020/04/habitat-destruction-covid19.

The same planes and ships: Tangley, "In Harm's Way"; Eichler et al., "Transmission of Severe Acute Respiratory"; Yu-Ching Chou et al., "A Major Outbreak of the COVID-19 on the *Diamond Princess* Cruise Ship: Estimation of the Basic Reproduction Number," *Journal of the Chinese Medical Association*

85, no. 12 (December 2022): 1145–53, https://doi.org/10.1097/JCMA.0000000000000820.

When SARS-CoV-2 emerged: Bedford, "Cryptic Transmission of Novel Coronavirus"; Elizondo et al., "SARS-CoV-2 Genomic Characterization"; Geoghegan et al., "Genomic Epidemiology Reveals Transmission"; Jean B. Nachega et al., "The Colliding Epidemics of COVID-19, Ebola, and Measles in the Democratic Republic of the Congo," *The Lancet Global Health* 8, no. 8 (August 2020): e991–92, https://doi.org/10.1016/S2214-109X(20)30281-3.

While controversy swirled: James C. Alwine et al., "A Critical Analysis of the Evidence for the SARS-CoV-2 Origin Hypotheses," *Journal of Virology* 97, no. 4 (2023): e00365-23, https://doi.org/10.1128/jvi.00365-23; Jesse D. Bloom et al., "Investigate the Origins of COVID-19," *Science* 372, no. 6543 (2021): 694, https://doi.org/10.1126/science.abj0016; The Lancet Microbe, "COVID-19 Origins: Plain Speaking Is Overdue," *The Lancet Microbe* 5, no. 8 (2024): 100953, https://doi.org/10.1016/j.lanmic.2024.07.016; Mark Woolhouse, "Did COVID Come from an Animal Market? Here's What the New Evidence Really Tells Us," *The Conversation*, September 25, 2024, https://theconversation.com/did-covid-come-from-an-animal-market-heres-what-the-new-evidence-really-tells-us-239533; Daniel et al., "Rapid Spread of African Swine Fever"; Spyros Lytras et al., "The Animal Origin of SARS-CoV-2," *Science* 373, no. 6558 (August 17, 2021): 968–70, doi:10.1126/science.abh0117; Wei Xia et al., "How One Pandemic Led to Another: Was African Swine Fever Virus (ASFV) the Disruption Contributing to Severe Acute Respiratory Syndrome Coronavirus 2 (SARS-CoV-2) Emergence?," Preprints.org, January 25, 2022, https://doi.org/10.20944/preprints202102.0590.v2.

Millions of pigs in China: Shibing You et al., "African Swine Fever Outbreaks in China Led to Gross Domestic Product and Economic Losses," *Nature Food* 2, no. 10 (September 27, 2021): 802–8, https://doi.org/10.1038/s43016-021-00362-1; "The $100-Billion Toll of a Pig Epidemic in China," *Nature*, October 1, 2021, https://www.nature.com/articles/d41586-021-02642-z; Lytras et al., "The Animal Origin"; Xia et al., "How One Pandemic Led"; Michael Worobey et al., "The Huanan Seafood Wholesale Market in Wuhan Was the Early Epicenter of the COVID-19 Pandemic," *Science* 377, no. 6609 (August 26, 2022): 951–59, https://doi.org/10.1126/science.abp8715.

We wanted uranium to move: Chelsea Gunter, "The Case for Uranium Mining in Greenland," *Cornell International Law Journal* 48, no. 2, article 5, https://scholarship.law.cornell.edu/cilj/vol48/iss2/5; Jim Robbins,

"A Nuclear Power Revival Is Sparking a Surge in Uranium Mining," *Yale Environment 360*, April 4, 2024, https://e360.yale.edu/features/us-uranium-mining-nuclear-power; Madeleine Gregory, "Greenland Has Climate Change-Fighting Minerals. Will It Mine Them?," *Atmos*, April 24, 2023, https://atmos.earth/greenland-rare-earth-elements-glaciers; Benjamin Ballinger et al., "The Vulnerability of Electric-Vehicle and Wind-Turbine Supply Chains to the Supply of Rare-Earth Elements in a 2-Degree Scenario," *Sustainable Production and Consumption* 22 (April 2020): 68–76, doi:10.1016/j.spc.2020.02.005.

Perhaps another pandemic: Eleftherios P. Diamandis, "The Mother of All Battles: Viruses vs Humans. Can Humans Avoid Extinction in 50–100 Years?," *Open Life Sciences* 17, no. 1 (January 29, 2022): 32–37, https://doi.org/10.1515/biol-2022–0005.

In about 250 million years: Alexander Farnsworth et al., "Climate Extremes Likely to Drive Land Mammal Extinction During Next Supercontinent Assembly," *Nature Geoscience* 16, no. 10 (October 2023): 901–8, https://doi.org/10.1038/s41561-023-01259-3; Rebecca Owen, "Future Supercontinent Will Be Inhospitable for Mammals," *Eos*, November 8, 2023, https://eos.org/articles/future-supercontinent-will-be-inhospitable-for-mammals; Elise Cutts, "Earth's Future Supercontinent May Be Too Hot for Most Mammals," *Science*, September 25, 2023, https://www.science.org/content/article/earth-s-future-supercontinent-may-be-too-hot-most-mammals; "New Research Reveals Extreme Heat Likely to Wipe Out Humans and Mammals in the Distant Future," *EurekAlert*, September 25, 2023, https://www.eurekalert.org/news-releases/1002455.

About the Author

Roberta Kwok is a science writer whose work has appeared in the *New York Times*, NewYorker.com, *Nature*, *New Scientist*, *Audubon*, and other publications. She has received a fellowship from the Knight Science Journalism Program at the Massachusetts Institute of Technology, a grant from the Alfred P. Sloan Foundation, and awards from the American Association for the Advancement of Science and the American Geophysical Union. Before becoming a journalist, Kwok worked in a genetics lab at Stanford University. She lives in the Seattle area and is originally from Canada.